The Race for Bandwidth: Understanding Data Transmission

Cary Lu

PUBLISHED BY
Microsoft Press
A Division of Microsoft Corporation
One Microsoft Way
Redmond, WA 98052-6399

Library of Congress Cataloging-in-Publication Data
Lu, Cary
The Race for Bandwidth: Understanding Data Transmission / Cary Lu
p. cm.
Includes index.
ISBN 1-57231-513-X
1. Signal theory (Telecommunication) 2. Broadband communication systems
3. Digital communications. I. Title.
TK5102.5.L79 1997
621.382'16--dc21 97-4204
CIP

Printed and bound in the United States of America.

2 3 4 5 6 7 8 9 QMQM 3 2 1 0 9 8

Distributed in Canada by ITP Nelson, a division of Thomson Canada Limited.

A CIP catalogue record for this book is available from the British Library.

Microsoft Press books are available through booksellers and distributors worldwide. For further information about international editions, contact your local Microsoft Corporation office or contact Microsoft Press International directly at fax (425) 936-7329. Visit our Web site at mspress.microsoft.com.

Acquisitions Editors: Eric Stroo, Anne Hamilton
Project Editor: Jenny Moss Benson

For the engineers of the digital age, with hope that
they will not see progress only as more,
or do things only because they can

For the children now and to come,
who will live out their analog humanity with the
consequences of this digital age

Table of Contents

Acknowledgments

Written on Cary's behalf by Ellen W. Chu

Cary was unable to write the acknowledgments for *The Race for Bandwidth* before he died, but he and I and our children owe a tremendous debt to many people for their professional and personal aid during the last year of his life while he tried to finish the book. As they have always done in our lives, the lines between the professional and the personal blurred in every case, and many of our colleagues are also close friends.

Our foremost thanks go to Stephen Manes and Adam Engst, without whom *The Race for Bandwidth* would not exist. Steve and his wife Susan Kocik, Adam and his wife Tonya Engst, Geoff Duncan, and Chris DeVoney kept Cary company during every hospital stay, talking technology and book bits all the while. Cary treasured their presence.

Thanks to the people and companies who provided information and equipment to Cary while he was researching this book. Unfortunately, I do not know their names and so can only apologize for failing to acknowledge them properly. Thanks to Cary's editors at Microsoft Press from years past and to the editors of *The Race for Bandwidth,* including Jenny Benson and Devra Hall and, especially, Acquisitions Editor Eric Stroo for his remarkable patience in seeing this project through from its beginning in 1996.

Thanks to Cary's brother Ponzy and the rest of the Lu family and to Cary's dearest friends and their families: Dan Aranovich, Marian Bremer, Dwight Davis, Katie Hafner,

Robert Kavanagh, Steve Kurtin, Steve Lambert, John Markoff, Ron McAdow, Robert Lin, Judy Peace, Jed Schwartz, Janet Shapero, Stephanie Tombrello, Allan Stone, and Joan Uglow. The assistance many of them gave as Cary lay dying was as indispensable as the years of friendship.

Thanks to my Chu family of physicians, who gave second opinions, moral support, and the kind of realistic advice Cary used to give them about computers. Deep appreciation to the staff of the University of Washington's Cancer Center and Medical Center Hospital, especially oncologists Stephen Petersdorf and James Bruckner, and of the Evergreen Community Hospice, whose efforts kept Cary working until days before his death.

For being here for Cary and me in our Northwest years, and for staying long after this book went to press, our gratitude goes to Jan Barrow and Stephen Miller, David Coder and Barbara Blankenship, Kristine Kaufman and Tom Bird, Deborah and Bruce Lajiness, Ted and Kathi Lucia, the staff and teachers at Carl Sandburg Elementary School, and everyone at Parkplace Book Company in Kirkland, Washington. The University of Washington's Department of Environmental Health offered me flexibility in my job throughout Cary's last year.

Thanks to Vicki and Daryl Cornell, who continue to be indispensable.

And to David and Marybeth Anderson, David Garrison, James Karr, Patricia and David Lincoln, Sharon Olds, Mark Pierson and Cathie Dunkel, and the staff of Northwest Environment Watch, who gave me what I needed when I needed it most.

Inexpressible gratitude goes to Jerre Levy and Lucia Roncalli, who for decades gave Cary what he needed whether he acknowledged it or not.

Finally, thanks and all our love to Meredith Lu and Nathaniel Chu, who provide so much more than the excuse for their parents' books being late. As Cary's prognosis became successively worse, he said over and over, "My only regret is that I won't see the kids grow up." I count myself unspeakably lucky that I am here to watch them now.

Foreword

by Ellen W. Chu

Life is what happens to you while you're busy making other plans.

— John Lennon, "Beautiful Boy (Darling Boy)," 1980

I picked up the phone. A clipped voice demanded, "Ellen Chu, please."

"Speaking," I said.

"I'm Cary Lu, managing editor of *High Technology* magazine. I want to know if you can teach a grown man to write."

So was I introduced to Cary Lu, author of this book and of the widely known *Apple Macintosh Book.* I was so mystified by that phone call that I searched the masthead of *High Technology,* published in Boston at the time, to see if this guy was for real. He was. But how had he heard of me, a biologist just out of graduate school and a new assistant professor of scientific writing at the Massachusetts Institute of Technology? I soon found out that professionally Cary hadn't heard of me and didn't really want me to teach any grown men to write. No; his mother had simply learned from my parents that I'd moved to Boston, and she wanted to set us up.

That was 1982; it took four years, but eventually we were married. Only fifteen years later, in September 1997, Cary was dead of cancer.

Teaching "grown men to write" was something Cary had had experience with. After his undergraduate training in physics and his graduate and postgraduate research in visual perception, Cary worked for a decade in film and television, notably producing science programs for *Nova, Sesame Street,* and *Infinity Factory.* During the seventies, a variety of science, film, and television consulting jobs had enabled him to travel around the world; live in Africa; and spend a year in a cheap New York City hotel so close to Lincoln Center that he could come home from work, nap briefly, and still take in an opera almost every evening. By the early eighties, Cary had decided he ought to find a more stable career. Since he was interested in technology, he bought a Zenith microcomputer and set out to "see if I could teach myself to write."

Cary wrote a lot during the following years. His style was as terse and distinctive as his phone manner. In my career as a science editor, Cary's was practically the only prose I wanted to *add* words to. Cary was as fanatical about accuracy as he was succinct; he also wanted every article and, later, every book he wrote to be the most comprehensive one on the subject. His 1982 article "Microcomputers: The Second Wave" for *High Technology* featured a 3-page, 24-column table covering 22 microcomputer models, including hardware specifications, manufacturers' addresses, footnotes, and extra explanations. A similar table for "Dawn of the Portable Computer" in 1983, covering 27 portables and "transportables" in another 3 pages and 29 columns, nearly drove the art department screaming out the door. To get numbers for these tables, Cary would walk the floors of industry trade shows with a magnifying glass, counting pixels on display screens and firing skeptical questions at the marketing staff.

Cary was a storehouse of obscure technical knowledge and fix-it know-how. His *Apple Macintosh Book,* which ran to four editions after it was first published in 1984, explained not only how to set up a Macintosh comfortably in your home or office, but also the xerographic technology found in copiers and laser printers, the basics of local area networks, and the complexities of data transmission between computers. Of the half-dozen computers still in our house, he was most proud of a PC he'd created from 99 cents' worth of parts and otherwise free stuff. His reputation for telling things straight became almost legendary in the microcomputer industry. When friends, even other journalists, wanted to know the full story behind a new product, they called Cary. Marveled one admirer, "Not only does he know everything, he always answers his own phone."

Cary loved technology and music and children. His Boston apartment was festooned with wiring, littered with audiovisual and computer equipment, and furnished with huge orange pillows left over from *Sesame Street* sets. A life-size cutout of Big Bird greeted visitors from the head of the stairs. Whimsical toys gathered dust on a shelf he'd built onto the wainscoting in a corner of his dining room. When we bought a house and moved to Seattle in 1986, he had the cellar remodeled into an electronic workshop and wired the whole house for Ethernet. Where the previous owners stored stemmed glassware, Cary stored hundreds of music CDs.

Yet throughout his career, Cary never bought into the technological fundamentalism of many computer, especially Internet, fans. Even as one of the architects of a digital age, Cary believed that computers and technology were tools, and like all tools, they could be misused. "Will one million Macintoshes sold mean ten million bad pictures?" he wrote in *The Apple Macintosh Book.* When the Lake Washington School District wanted to pass a couple of technology bonds in the mid-1990s, Cary and I tried hard to convince the principal of our children's elementary school that it was much easier for a child to avoid learning to read than to avoid learning how to use a computer. "The last time a big technology levy passed, you installed a bunch of Commodore 64s that nobody used. Now you have a bunch of Apple IIs. Didn't the district learn anything from those experiences?" After we lost that battle, and the district won its levy, Cary sold the school a couple of Macintoshes—and a couple of fine microscopes—and then devoted considerable time to helping the staff build and maintain a working, useful computer system.

When Cary first told me his plans for a book about bandwidth, my reaction was, "What's bandwidth?" Only after reading the galley proofs did I realize that the topic brought together all his childlike fascination with, and entirely adult expertise in, technologies across the bandwidth spectrum. As I read Chapter 2, "A Brief History of Communications," I recalled a car trip we took along the Outer Banks of North Carolina in 1985. Required stops were windy Kitty Hawk, where the Wright brothers launched their first successful flight, and the beach from which Guglielmo Marconi sent the first transatlantic wireless message. It would not surprise me if Cary had written that whole chapter off the top of his head.

Though he died before he could finish *The Race for Bandwidth,* the text is still quintessential Cary. All but two of the chapters had been largely written, and he had given a lecture that laid the groundwork for the Internet chapter. In his dying weeks, he spent many hours discussing ideas for still unwritten sections with his close friends, Stephen Manes, who writes columns on computers for the *New York Times* and other publications, and Adam Engst, publisher of the electronic newsletter *TidBITS* and author of the *Internet Starter Kit* series. They promised to complete the book, and I am grateful that they have. In addition to crafting the Internet chapter from Cary's lecture notes and updating several sections concerning technologies that moved ahead after his death, Adam and Steve have thoroughly reorganized and edited the entire text. Without them, there might be content, but there wouldn't be a *book.*

A decade ago, scrambling to finish the revisions to the third edition of *The Apple Macintosh Book,* Cary and I paid wry tribute to our daughter Meredith, "who arrived before her parents could complete the revisions and stayed long after the book went to press." The acknowledgments to the fourth edition read, "The last edition was delayed by the arrival of Meredith; this one was delayed by Nathaniel. For the next edition, I will have to think of a different excuse." Now it is *The Race for Bandwidth* that is late, because life keeps happening as we make other plans.

Kirkland, Washington
May 1998

Chapter One

Why Bandwidth Is Crucial

Our information society revolves around the idea of bandwidth. The power and profits of telephone companies, television broadcasters, cable companies, and Internet service providers fundamentally lie in their ability to control—or attempt to control—the way bandwidth is deployed and used.

Governments consider bandwidth of the utmost importance. Bodies such as the Federal Communications Commission and state public utilities commissions have been set up specifically to regulate bandwidth through telecommunications acts, spectrum auctions, and official tariffs. Often the most important issues are in subtle details that may not be noticed when buried deep in legislation, even when broader concerns are debated widely.

Bandwidth is so important that governments insist on regulating it

What Bandwidth Means

Just what does this strange thing called "bandwidth" really mean? Bandwidth is a measure of how much information can flow from one place to another in a given amount of time. That information flow can take many forms—a telephone conversation over a wire, a television program broadcast over the airwaves, or a CD-ROM in your computer

Bandwidth is a measurement of information flow

sending data no farther than the screen. But the term has increasingly come to refer to the capacity of the transmission medium or device, as in the question "How much bandwidth will your modem connection give you?" The more information you want to send over a given time, or the faster you want to send it, the more bandwidth you need. So this rather simple technical notion has very far-reaching implications.

Higher bandwidth can deliver more information

Take a simple audio example: although telephone calls are quite intelligible, they are nowhere near as clear as what you hear on FM radio. That's mostly because an FM radio broadcast uses six times the bandwidth of a single telephone call. That extra bandwidth helps FM deliver higher high frequencies, lower lows, and less noise.

Figure 1-1

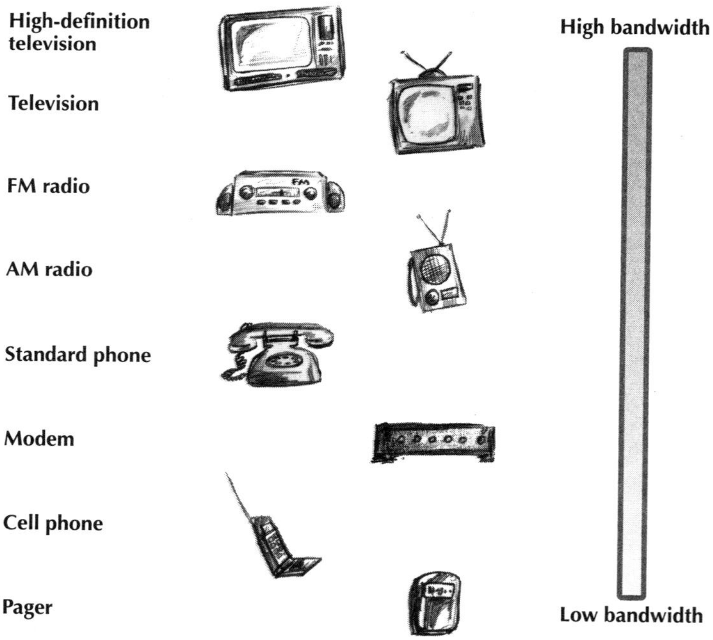

Now consider a question involving video: Why do movie clips from a CD-ROM usually look small, jerky, and fuzzy compared with what you see on television? The answer is that CD-ROM provides just a small fraction of the bandwidth of broadcast television. To put things into perspective: broadcasting a single high-definition TV channel requires the bandwidth of roughly 300 phone calls.

Figure 1-2

Video requires far more bandwidth than audio.

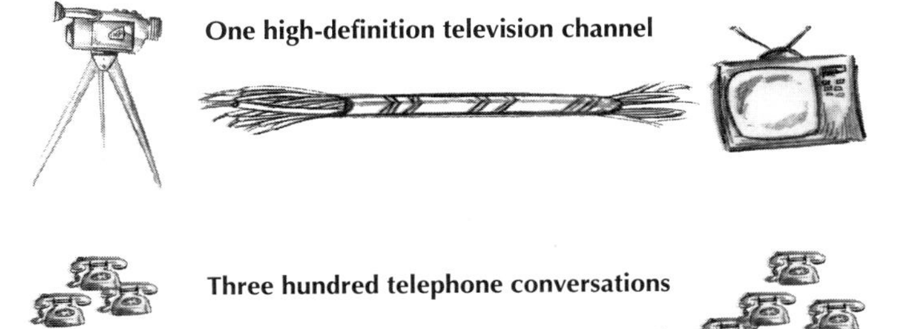

Why You Must Understand Bandwidth

Bandwidth affects the way we experience content

Why bother thinking about bandwidth? For one thing, as we've just seen, it directly affects the way we experience content. Imagine what would happen if television's bandwidth were suddenly reduced by a factor of roughly 200—the result would be virtually unwatchable. In fact, these days, you don't even have to imagine it: just take a look at any of the miserable videos you can get over a low-bandwidth modem connection to the Internet.

The availability of bandwidth affects the way companies conduct business

Bandwidth is the lifeblood of any business. Imagine trying to work without telephone, fax, or electronic mail. Now project that just a few years out, and you'll begin to realize that the decisions your company makes about bandwidth will play a key role in its success or failure. Will you be connected to the rest of the electronic world every minute of the day through a high-speed Internet link, or will you have to make do with an intermittent modem connection? When you attend teleconferences, will faces and voices be hard to make out, or will they seem as clear as if your colleagues were in the room? The answers are largely matters of bandwidth.

Access to bandwidth affects people's leisure activities

Nowadays most leisure activities also depend on bandwidth. Will you soon be watching movies and sports at home on big high-definition screens over channels that require even more bandwidth than those we have today, or will broadcasters instead chop up that same bandwidth into two dozen channels that look slightly worse than the ones we have now? Will our World Wide Web connections be able to deliver full-motion video that looks wonderful, or will we still be calling it the World Wide Wait? The answers are all about bandwidth.

Seeds of Confusion

Facts and figures about bandwidth can be misleading

Understanding the basics of bandwidth is not terribly complicated. The complex problem is separating hype from facts. Today, many announcements and articles about bandwidth are filled with errors, misunderstandings, and deliberately misleading statements. Often you can't tell if the person making a misstatement is trying to mislead or is simply confused. The technology industries have not helped, routinely throwing around inaccurate or simplistic numbers and concepts. For example, as you'll see in Chapter 8, "Bandwidth and the Internet," a so-called 10-megabit Ethernet connection never delivers data at 10 megabits per second; it doesn't even come close. And you'll also see why digital is not always better than analog, despite what some people would have you believe.

The Struggle to Control Bandwidth

Why are so many large companies in the publishing, broadcasting, computing, and telecommunications industries merging, making alliances, and angling for public visibility? They are maneuvering to control bandwidth. At one level, these companies want simply to sell the connections, just as telephone companies now sell their services without influencing what you say or hear over the phone. But at another level, these companies want to control content as publishers; they want to generate and get paid for the information you see and hear.

Companies now want to control both the content and the means of delivering it

In the past, the companies that supplied the connections were rarely the same ones that supplied the information. Today, these roles are blurring. The major players are now acting more like cable television companies. Cable companies control both their connections and their content, picking and choosing the channels you can receive on the basis of popularity—and which channels they happen to have investments in. When you combine control of the pipeline with the information that flows over that connection, the result is leverage that can be applied to increase profits or even manipulate public opinion.

Combined control can increase profits and influence public opinion

The Internet Changes the Rules

The Internet, and in particular the World Wide Web, offer a way to bypass traditional publication and distribution. For the first time, the ability to gain access to bandwidth is not the exclusive provenance of major companies or government agencies. On the Web, virtually anyone who really wants to can publish content for anyone else who wants to see it. Today on the World Wide Web, any page is essentially as easy to reach as any other, and it typically costs the same to view: nothing.

The Web levels the playing field, providing a forum for anyone with access

Control of bandwidth will affect user access

Depending on who prevails in the control of bandwidth, the nature of the Internet may change considerably. Internet providers may make it easier and faster for you to access pages sponsored by publishers and advertisers who have paid for the privilege than to access pages developed by your Aunt Elma. Things are already changing: pricing is beginning to vary depending on the nature of the information you want and whether you have to look at an advertisement to get it.

Informed decisions are at the heart of a democratic country. Just how well informed will we be? That will depend not only on all of us as information sources and information consumers but also on who controls bandwidth and how it is made available to us.

A Limited Resource

The supply of bandwidth is finite

Bandwidth will always be a limited resource. That has long been obvious for broadcasts such as radio and television over airwaves. The available frequencies are not unlimited, and two stations cannot use the same frequency in the same area at the same time without **interference**. Those limitations of broadcast media led directly to governmental regulation.

Increasing bandwidth is expensive

Bandwidth supplied by wires—telephone, cable TV, local area networks—may seem less limited, because wires can always be added or replaced by higher-capacity connections. But the high cost of installing new wires or fiber-optic cables and the difficulty of integrating new capacity with old systems place functional limits on the available bandwidth for years and even decades.

The Drive to Increase Bandwidth

Because bandwidth is so valuable, there are tremendous incentives to increase it. And as you might expect, most of the players in the information society—from program producers to computer designers to telecommunications companies—are trying their hardest to do just that. Even with the available communication paths, technical tricks can help. Color television, for example, was shoehorned into a bandwidth that was originally designed for black-and-white television, and digital compression has helped videophones deliver color pictures over lines originally meant for voice communication.

Maximizing the use of available bandwidth is the first trick that is tried

But each time the clever tricks run out, a new communications path with higher bandwidth is proposed as a new standard. High-definition television, for example, is intended to replace our present television system, but like most new standards, it will not be cheap to implement. High-definition television requires new hardware and software every step of the way—new cameras, new transmitters, new television sets, and new VCRs, and on the software side, new programming developed in the new format.

When available bandwidth is used to maximum capacity, new standards and technologies are born

In the past, the improvement has sometimes been so compelling that a new standard takes over quickly, such as when the compact disc replaced the vinyl long-playing record. For high-definition television, however, arguments over the costs, technology, and bandwidth required have delayed adoption for more than a decade...so far.

Sometimes new standards are slow to be adopted

But not everyone is rushing to increase bandwidth. Some companies are making so much money from their present bandwidth structure that they are reluctant to develop higher bandwidth offerings at what would be inevitably at a lower price. Other companies seek to delay industry-wide bandwidth improvements until they can get their own

Keeping resources limited can be valuable to those who own and control them offerings ready. And some companies are behaving like the telegraph companies that tried to ignore the introduction and proliferation of the telephone.

Costs Drop, Rules Change

Bandwidth has a history of declining prices

The overall cost of bandwidth has dropped significantly over the years. Many people (sometimes known as parents and grandparents) who lived through the olden days of telecommunications rush to finish long-distance phone calls, behaving as if the calls cost more than $3 a minute, as they did in 1945, instead of less than 15 cents, as they typically do today. Bandwidth has a history of declining prices, and that history is certain to continue. But the price of bandwidth is very uneven: for the same bandwidth, some people pay nothing and others pay quite a lot. Governmental regulations enforce these disparities in the interest of some greater good—like having affordable service for citizens in rural areas.

Barriers Fall

As prices drop, access and diversity of content increase

As the cost of bandwidth drops, so do the barriers to publishing and distributing information. Forty years ago, only a few producers could get programs on television, and large publishers controlled the flow of books into bookstores. Now, new television networks and cable channels emerge weekly to take advantage of expanded cable systems, and the World Wide Web permits a vast new group of "information providers" to publish Web pages at very little cost. Lower costs mean that many more points of view can be heard, points of view that might not have been supported by traditional publishers.

As content sources mushroom, consumers must learn to discern their value

Many pages on the Web consist of personal material that in decades past might have been confined to a diary or perhaps sent to one or two friends. When publication was difficult and expensive, publications were more likely to have been prepared with care and effort. Now, with the barriers to publication so low, sloppily prepared materials without editing, fact-checking, or careful thought are more common than ever. The explosion of information puts a burden on the reader, who must put more effort into separating worthwhile information from the irrelevant or worthless.

Will Bad Information Drive Out Good?

Free or very cheap information over the Internet threatens to replace more traditional media that cost money, such as newspapers. Because the quality of information over the Internet is at best uneven, this could turn into a Gresham's Law of Information, with bad (cheap) information driving out good (expensive) information. Are people willing to pay for high-quality information over the Internet? Can such information be protected from piracy? For now, no one knows.

People may not wish to pay for content even if it is of superior quality

Quantity, Not Quality

Bandwidth is simply a measure of the rate of flow of information; it does not distinguish between high-quality and low-quality information. And bandwidth doesn't measure the amount of information actually understood and absorbed. A piece of junk mail and a personal letter use the same bandwidth, whether on paper or in electronic form, and every medium now has its equivalent of junk mail. Moreover, each medium also seems to have its junkies—people who cycle endlessly through cable infomercials or stay glued to online chats. Some enthusiasm seems to be based only on media novelty, such as the electronic services that seek out the "worst" pages on the Web. Does anyone search out the worst junk mail on paper?

Bandwidth does not recognize quality

Jargon and Reality

Casual use of the term "bandwidth" may confuse the issues

In pop techno-jargon, bandwidth has taken on informal meanings. A "high-bandwidth" conversation or meeting is filled with content—or at least facts. Calling a person "low-bandwidth" is a way to put down someone who is slow in thought and conversation. But remember that bandwidth does not respect quality of information. A fast-talking salesperson might offer apparently high bandwidth with little useful or accurate content; a person of few words might choose them more judiciously.

Bandwidth Increases, Time Does Not

Increasing bandwidth will deliver more data, but will you have time to use it?

Although bandwidth can increase, time does not. What's the point of increasing bandwidth to deliver more information to our homes and offices when we can't cope with what is arriving already? Increasingly, we can find time to pay attention to a new form of bandwidth, such as CD-ROMs or the Internet, only by paying less attention to an older form, such as television. And the older media keep expanding: half a dozen channels of broadcast television have given way to 300-channel cable TV systems and direct broadcast satellites.

Bandwidth continues to increase, but people are working longer than ever

Proponents point out that greater bandwidth can save time: telecommuting from home can minimize time spent in traffic, and teleconferencing can save a trip to another city. But the technology costs money and may crash at the most inconvenient times. And if you can have an office at home, you can never completely leave your office. Do new technologies that increase bandwidth save more time than they consume? Teleconferencing may eliminate a trip, but if you spend days trying to get the equipment working or if you can't operate as efficiently over the teleconferencing connection, the overall gain may be negligible. Surveys show that the average American has worked longer hours over the past four decades, with a corresponding reduction in leisure time. Yet the amount of bandwidth available has increased tremendously in that period.

Some Modest Predictions

This book looks primarily at our current situation and ahead to the next five years or so. The reason is simple: anything that will be available in the year 2003 is almost certainly available in working prototype today. In 2003, there are unlikely to be any magic technologies in widespread use that we don't know about today. On the other hand, high-definition television and videoconferencing over the Internet are technologies that are not in widespread use today, but they will certainly be with us in 2003 and quite possibly sooner.

Innovations for the near future are already in development

If It's Not Here Now . . .

Conversely, if something doesn't work at all today, chances are that it won't be working in 2003 either. True video on demand, where every movie and TV show ever made is available instantly at your TV set whenever you want it, is almost certain not to happen. Intelligent agents, software that does your bidding in a humanlike way, are also highly improbable in the short run. Despite much talk about agents, no truly useful intelligent agent has been demonstrated today, so it's very unlikely that they'll evolve by 2003. The only intelligent agents I know of are people, and I do expect to see people selling their services as agents to help cut through the information glut.

Many innovations now being promised may never see the light of day

Speaking of Business

Predicting available technologies is the easy part. Business issues such as the costs of technologies and which companies might benefit from them are much harder to guess. This book will indulge in some business predictions for the near term, but they are guaranteed to be shakier than the technology discussions.

What Do You Want? What Will You Pay?

Technophiles represent only a small segment of the public at large

A third kind of prediction is harder still. What does the public want from the Internet and other sources of information? What are people willing to pay for, and how much will they pay for it? These questions will be decided by millions of people around the country and around the world. In a very real sense the readers of this book, collectively, are in a better position to judge these issues than the author. But bear in mind that if you're reading this book, you're part of a self-selected group, interested enough in technology to spend time learning about it. That immediately puts you and your fellow readers in a rather small group that's almost certainly not representative of the public as a whole.

Keep in mind that the public never asked for radio, television, or video games. Similarly, asking someone who has never seen the World Wide Web about what kind of Web access might make sense for him or her is probably not a very useful exercise.

Will the Internet Ever Be Fast Enough?

For the millions of Web users, the most eagerly awaited bandwidth improvements relate to the Internet. Many Internet users have focused on increasing the speed of the one segment that they have control over—the final link from their Internet service provider to their home or office—by upgrading modems or other connections.

Unfortunately, this is yet another bandwidth perplexity: the "final mile"—the connection that actually terminates at your home or business—is only one of many links that Internet information passes through, so a fast final connection doesn't speed up everything. Internet usage is increasing so dramatically, both in the number of users and in the variety of information available, that some people have argued that the Internet will begin to collapse from the traffic. Or if the Internet doesn't collapse, it will never get much faster because the backbone—the main long-haul trunks that carry the heaviest data loads—might not be able to be expanded fast enough to outrun the demand. Internet bandwidth is the most interesting—and complicated—form of bandwidth today. Because it relies on so many general principles of bandwidth, I will discuss it mostly in Chapter 8, "Bandwidth and the Internet." Feel free to skip ahead if the Internet is your primary concern.

The end-user's control over bandwidth is limited

Chapter Two

A Brief History of Communications

The history of bandwidth has been a constant quest for more—a quest that continues to this day. People have always wanted to communicate more information in less time, and the likelihood is that they may never be entirely satisfied. The more information you can transmit and receive, it seems, the more you want.

Telecommunications

Distance is one of the greatest challenges

Sending information over long distances has been a challenge for most of human history. Runners and riders were once the only practical methods for communicating detailed information. Drums and smoke signals were too limited for any but the simplest messages. There was no lack of trying, however; elaborate arrays of smoke signals or signal fires that could encode text were tried thousands of years ago, but the naked eye could not see far enough to make those methods very useful.

Figure 2-1 ***Early methods of sending information over long distances were limited and slow.***

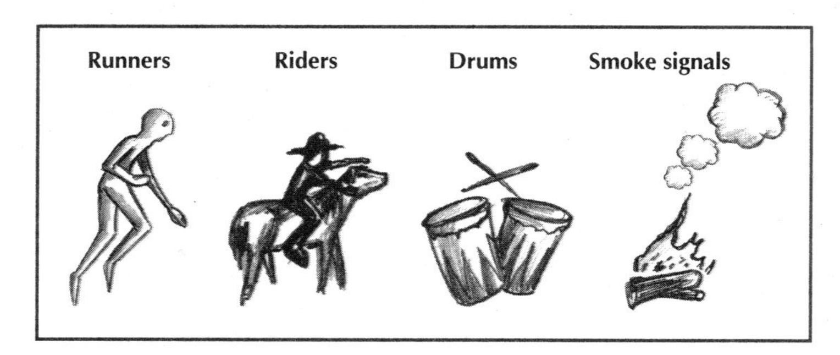

Telegraph

Optical telegraphs first used a semaphore method

The invention of the telescope made fast long-distance communication possible because signals could be seen from a much greater distance than with the naked eye. Starting at the end of the eighteenth century, the optical telegraph was widely installed throughout many countries. Optical telegraphs first used a semaphore method with two wooden arms that could be rotated to seven positions each, allowing 49 alphabetic characters and symbols to be encoded. A different system using shutters could create black and white grid patterns capable of encoding 64 characters. These early telegraphs looked something like those shown in Figure 2-2.

Figure 2-2 ***Optical telegraphs were the first ways of moving data quickly across long distances.***

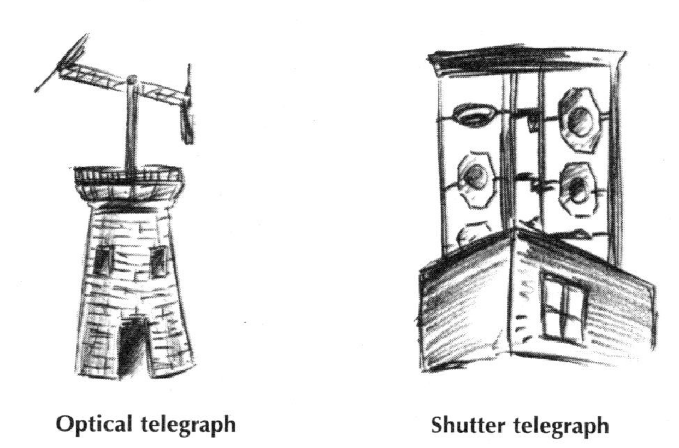

The optical telegraph ran along a series of relay stations where attendants copied and repeated messages. For the best sight lines, the stations were located on hilltops, hence the many little peaks called "Telegraph Hill" or "Station Hill" around the country. Data transmission speeds depended on the number of relays. For example, an early French system from Toulon to Paris spanned 480 miles with 140 relays and could send 50 characters in 40 minutes. Because each of the 140 towers required an attendant, the system was expensive; its use was limited to the government and to very wealthy businesses.

Optical telegraph transmissions were expensive, and speed depended on the number of relays

The first practical electrical telegraph, the line from Baltimore to Washington, D.C., built by Samuel Morse in 1844, quickly made the optical telegraph obsolete. Now signals traveled along iron wires strung on telegraph poles, often alongside a road known as "Telegraph Avenue." The original scheme used electromagnetic markers to record the signal on a strip of paper, but soon telegraphers simply listened to clicks corresponding to the binary dots and dashes of Morse code. Speeds of 25 words per minute were normal.

The average rate of electrical telegraph transmissions was 25 words per minute

Telegraphy expanded rapidly, beginning in North America and Europe. In 1866, the first telegraph cable was successfully laid across the Atlantic Ocean. Newspaper publishers and stockbrokers installed private telegraph lines, which began the practice of leased or dedicated lines. The telegraph evolved into telex services with a typewriter-like keyboard replacing the telegraph key. Telex remains in use today, but it has been largely replaced by fax and electronic mail, both of which offer significantly higher quality and speed, along with far greater availability. Although 20 telegraph cables eventually crossed the Atlantic, the last purely telegraphic transatlantic cable went out of service in 1966, exactly one hundred years after the first cable was laid.

Enhancements to telegraphy led to the telex, which is still in use today

Telephone

Separating telegraph signals into different audio frequencies led to the invention of the telephone

Telegraphy went through many improvements, particularly to increase speed and develop methods for carrying multiple data streams on a single line—both efforts to achieve higher bandwidth. One way to provide multiple data streams was to separate the signals into different audio frequencies, a technique that is used today in modems. Alexander Graham Bell was working on this technique when he extended the idea to make the telephone, producing the first commercial system in 1877. By 1880, there were 30,000 telephone subscribers.

Telephone technology requires more complex switching than the telegraph

The telegraph typically involved only one central telegraph office in each town, so switching was simple—the operator had only to decide which telegraph line ran toward the destination. The telephone, on the other hand, requires complex switching, because any telephone can be connected to any other telephone. At the turn of the century, the rapid growth in the use of the telephones and the need for operators to manage the increasingly complex manual switches led to predictions that within a few decades every adult would have to be pressed into service as an operator.

This prediction has, of course, come true; we are all operators now as we dial our calls. The first fully automatic telephone switch was installed in 1921. Direct long-distance dialing became common four decades later within the United States, and internationally in 1971.

Vacuum tube amplifiers helped to expand distance boundaries

An early telephone call could travel about a thousand miles before the signal became inaudible. Starting around 1915, vacuum tube amplifiers erased distance limits; three relay amplifiers were sufficient for a transcontinental call. After World War II, microwave relay towers for long-distance calls largely replaced telephone cables over land. Fiber-optic cables replaced these in the late 1980s.

Crossing oceans was much more difficult than crossing continents, because for many years, amplifiers were not reliable or small enough to install in a cable (the simpler telegraph cable required no amplifiers). Instead, transatlantic telephone service started via short-wave radio in 1926. Thirty years later, the first telephone cable across the Atlantic was laid; it could carry just 30 simultaneous telephone conversations. Because microwave transmissions follow a straight line and cannot cross an ocean, the next improvement was the use of satellites, which started carrying transatlantic phone traffic in 1963. But sending the signals the very long distance to a satellite in a stationary orbit inevitably caused short pauses that made the calls awkward. The first transatlantic fiber-optic cable, laid in 1988, largely replaced satellites for phone calls; that cable can carry 37,800 simultaneous conversations.

Telephone cables gave way to microwave relay towers and ultimately to fiber-optic cables

Cellular phone service began in 1983 and has grown so rapidly that some people predict that the majority of voice calls will be wireless in the near future, a bold but uncertain prediction that will rely in large part on the cost structures that evolve. For instance, in some countries, the caller always pays for the call, whether or not it's received on a normal phone or a cellular phone. In those places, notably Israel and Australia, cellular phone penetration is extremely high. Although cellular phone service is almost always of lower quality than normal telephone service due to its limited bandwidth, the convenience afforded by a wireless telephone outweighs the quality tradeoff.

Fax

Arthur Korn developed the first practical fax (facsimile) machine in 1902, sending pictures over a wire connection. In 1913, the portable Berlino fax machine enabled fax images to travel by telephone lines. For many decades, only fax machines from the same manufacturer could

Early faxing used proprietary standards

communicate with one another, and each manufacturer had its own proprietary protocols.

Enhanced facsimile standards enabled transmissions of higher quality in less time

Finally in 1974, a United Nations agency set the first international fax standard, Group 1, which allowed one page to be sent in six minutes at a resolution of 98 dots per inch. However, fax did not begin to become widespread until 1980, when the Group 3 standard was adopted, enabling one 200-dots-per-inch page to be sent in one minute, and later in 30 seconds.

Radio and Television

Radio meant wireless communication

At the turn of the century, radio emerged as a way to send telegraph messages in Morse code without wires. Transatlantic radio transmission was demonstrated as early as 1901 by Guglielmo Marconi, and the first voice transmissions by radio were made in 1906 by Reginald Fessenden. The advent of commercial radio broadcasting in the 1920s quickly turned radio into a new mass medium. Short-wave broadcasts routinely spanned oceans starting in the 1920s. FM radio transmissions, with far higher bandwidth and thus better sound quality than the older AM radio, began in 1939 with inventor Edwin Armstrong's experimental station; stereo FM commenced in 1961.

Radio and television transmission frequencies are subject to government regulation

Television broadcasts started on a limited basis just before World War II and grew spectacularly after the war with the first transcontinental distribution in 1951. Color television started in 1953, but did not become popular until the mid-1960s. Radio and television have been subject to close government regulation because they use scarce spectrum space and will interfere with each other unless their frequencies and broadcast antenna locations are carefully planned.

Transmitting television signals across the Atlantic was a much more challenging problem than transatlantic radio had been. Slow transmission experiments in the late 1950s (one minute of television was sent as a twenty-minute telephone call) gave way to satellite relays in 1964.

Satellite relays were used for television in the mid-1960s

Cable television started as community antenna systems, where a neighborhood far from local stations or in hilly terrain with poor reception would use a tall tower to pull in signals and send them via coaxial cable to each subscriber. In the 1960s, cable television companies began appearing in areas where reception was not an issue. This newer form of community antenna system offered additional programming that could not be received from nearby broadcast stations.

Cable television has expanded from its small-town roots

By the 1980s, more than 70 percent of the U.S. population was within reach of cable systems, and these systems offered from 25 to 150 channels of programming that could accommodate niche market tastes far better than the broadcast spectrum could. Large satellite antenna dishes proliferated around the world, put up by viewers who wanted more programming choices, initially in areas that cable television did not serve. By the 1990s, special satellite broadcasts aimed at small antenna dishes began competing directly with cable.

Satellite antenna dishes provided even greater choice and flexibility than cable

Communicating with Computers

Although the telegraph (optical and electrical) and fax machines are forms of digital communication, the real development of data communication began with digital computers. The fundamental measure of digital bandwidth is bits per second, abbreviated as bps. The abbreviation for kilobits per second, or thousands of bits per second, is Kbps. Mega or millions of bits per second is Mbps, and Gbps is giga or billions of bits per second (in American usage; milliards in British usage). These definitions will be explained more fully in the next chapter.

Digital bandwidth is measured in bits per second (bps)

If telegraph transmission speed had been measured in bits per second, the data transmission speed of the French optical telegraph system described earlier in this chapter would have been about one tenth of one bit per second. The speed of 25 words per minute for the electrical telegraph is equivalent to 15 bps. Compare this to the telex or teletype links for transmitting information to and from a computer that first appeared in 1940 and remained common until about 1970; their speed was typically 110 bps. Bell Telephone Laboratories introduced a 300-bps modem standard in 1958. In 1964, the American Standard Association promulgated **ASCII** (American Standard Code for Information Interchange) coding, enabling all standard English characters and punctuation to be represented by seven bits of information.

Using standard telephone services, today's high-speed modems support 33.6 Kbps

The Carterfone of 1966, an acoustic coupler and modem that could transmit data over a standard telephone handset, represented another milestone. After a long legal battle, it broke the Bell Telephone Company's monopoly on equipment that could be attached to a telephone line, leading the way to rapid development of faster and cheaper modems as shown in Figure 2-3. As a result, 1200-bps modems arrived in 1980, and higher speeds followed in succession: 2400 bps in 1984, 14.4 Kbps in 1991, 28.8 Kbps in 1994, and 33.6 Kbps in 1996.

Figure 2-3 ***Modem speeds over dial-up phone lines have increased rapidly.***

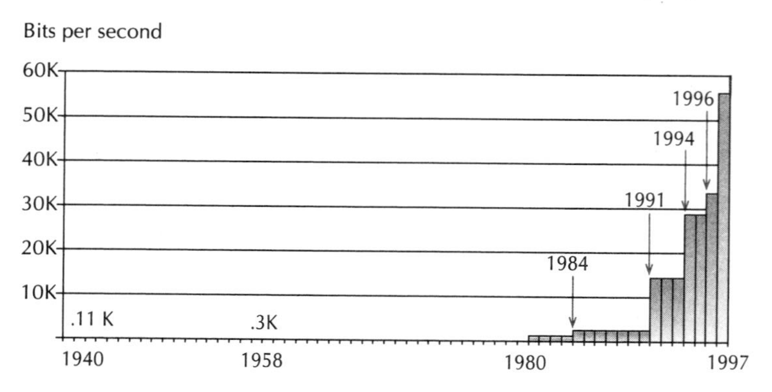

For standard telephone service, 33.6 Kbps is close to the speed limit imposed by the switch and signal relay circuits, though not by the copper phone line wires. However, so-called 56K modems, which can receive data at a maximum speed of about 53 Kbps but use the 33.6 standard for transmitting, began appearing in 1997. They gain their extra speed from the avoidance of an analog-to-digital conversion by the device that sends data to them.

Avoiding analog-to-digital conversion can enhance speed

ISDN (Integrated Services Digital Network), defined in 1984 and introduced in the United States in 1992, can run at 128 Kbps over the same copper wires.[1] Higher speed data connections over ordinary telephone and cable television lines promise speeds well over 1000 Kbps.

ISDN phone service uses a maximum rate of 128 Kbps

Computer Networks

All of the previous methods of data communication were designed to connect one computer to another, much as one person calls another via the telephone. However, data communication turns out to be far more useful and powerful when numerous computers are all connected together to form a network than when each one makes a direct connection with another.

Networks erased the boundaries of one-to-one communication

In 1964, the Advanced Research Project Agency of the U.S. Department of Defense developed a network called ARPANET, the precursor to the Internet, to connect the small number of then-powerful computers located at important research sites around the country. Other networks sprang up in the 1980s, but the Internet gradually absorbed them, being in essence a "network of networks." The Internet itself became more widely used in the mid-1990s, and today computers routinely

ARPANET was the precursor to the Internet

1. For more information about the history of ISDN, read *The Golden Splice: Beginning a Global Digital Phone Network*, a Web page written by John Landwehr of Northwestern University at *http://www.tiug.org/whatis.html.*

use it to communicate with one another. Figure 2-4 shows the tremendous growth in the number of host computers over the years.

Figure 2-4 ***From 1969 to 1998 the number of host computers connected to the Internet grew from 4 individual hosts to almost 30 million.***

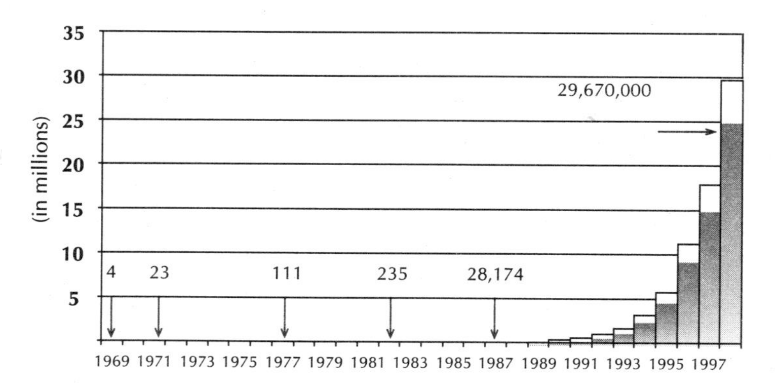

Modern local area networks can run at 1000 Mbps

As the prices of computers dropped and their numbers increased, the drive to connect all the computers in a company or office gave rise to local area networks. When data and hardware resources such as printers are shared across local area networks, faster is always better. Xerox introduced Ethernet, the most popular type of local area network (LAN), in 1975, with a nominal speed of 10 Mbps. Several faster versions of Ethernet have grown in popularity, including a 100 Mbps version introduced in 1993. Other types of networks use fiber optics and other protocols that run at over 150 Mbps, and the so-called Gigabit Ethernet is 1000 Mbps.[2]

---

2. For more information on the Gigabit Ethernet, visit the Web site of the Gigabit Ethernet Alliance at *http://www.gigabit-ethernet.org/*.

Packaged Media

One final form of communication is worth mentioning: media that deliver information largely in packaged form rather than in a stream of data. The major packaged medium for information for the past four centuries, of course, has been ink on paper in the form of books, magazines and newspapers. Improvements in printing technologies have enabled volumes to increase and prices to drop and have added new forms of information such as photographs. And the book promises to be the medium that will outlive all other packaged media. It is the only form that will surely be readable in another century (if printed on acid-free paper), at a time when virtually no one is likely to have the hardware necessary to read a CD-ROM as we know it today.

CD-ROMs, floppy disks, and other media provide an alternative to transmitting data in a stream

Thomas Edison's cylinder phonograph began another packaged medium, this one for the delivery of sound. Invented in 1877, it was largely replaced by disc phonographs around the turn of the century. The early discs (called 78s because they rotated at 78 rpm, or revolutions per minute) were supplanted in turn by the long-playing 33 rpm record (which arrived in 1948 and was enhanced by stereo in 1958) and the compact disc (which first appeared in 1982). Magnetic recording, developed initially as a wire recorder by Valdemar Poulson around 1900, began to be replaced by reel-to-reel magnetic tape in the late 1930s. The first cassette recorders arrived in 1962; Digital Audio Tape (DAT) in 1986.

Audio recording technologies evolved from analog to digital

Motion pictures were originally delivered on various media, including flip-cards, but 35-millimeter films began in 1888. They remain with us today in nearly the same form, albeit with color and surround sound added. But today motion pictures are often viewed on videotape, first introduced for studio use in 1957 and then in cassettes as a home television

Video recording followed a similar path of evolution

format; the common VHS version arrived in 1976.[3] Laserdiscs were introduced in 1978 and became commercially available in 1980; the DVD format first appeared in 1997.

Computer media such as floppy disks and CD-ROMs are in their infancy, but they, too, deliver data in a compact package. As you will see in the next chapter, packaged media can be thought of as having a bandwidth of their own, because they deliver a certain amount of information over a certain amount of time and distance.

In Conclusion

Each innovation was, in effect, the result of increasing bandwidth

Even this very brief history demonstrates the importance of communication to society. The incentive to improve bandwidth in order to send information faster and farther (and cheaper) propelled each new innovation. Yet until the early part of this century, businesses and governments were the main users of private messaging services such as the telephone and telegraph. Today, individuals who use communications services and information at home and in small offices are just as much a motivating force in the development of bandwidth and the Internet as are larger institutions. In the next chapter, you'll learn the basics of what bandwidth is all about.

---

3. For a more detailed history of the VCR, see *The History and Technology of the VCR* at *http://www.eia.org/cema/prod-hut/hometh/files/hstryvcr.htm*.

Chapter Three

Thinking About Bandwidth

Bandwidth is a measure of the information that can flow from one place to another in a given amount of time. The issue isn't merely the speed of transfer of a particular piece of information; it's the total amount of information moved in a fixed amount of time.

Width is more important than speed

That's why it's essential to think about the "width" of the communication channel. A wide, slowly flowing river often moves much more water than a rushing stream, because the width of the river means that more water can flow along at any given time. In the world of electronics, signals typically travel at the same extremely high speed (usually somewhere near 186,000 miles per second, or the speed of light), so distances and speeds tend to matter less than conceptual "width"—how much information can travel simultaneously.

For short distances or small amounts of data, the impact of bandwidth may be almost unnoticeable. But as distance and amount of information increase, bandwidth becomes more important. And, surprisingly, the best solution may not be the electronic one.

Where Did the Name "Bandwidth" Come From?

The term is an artifact of the terminology of radio, where the electromagnetic spectrum was visualized as and divided up into "bands" such AM, FM, and various short-wave bands. The bands themselves and channels within them have a "width" expressed in some multiple of cycles per second (**hertz**). Within limits, the wider the band, the more information it can broadcast. The primary reason AM radio sounds so much worse than FM is bandwidth: each FM station gets 200 kilohertz (200,000 cycles per second) of bandwidth for itself, whereas an AM station gets just 10 kHz. The difference is clearly audible.

A Slow Boat Beats a Fast Wire

A European T1 line can transfer information from New York to London at approximately 2.048 Mb per second

To see just how important "width" can be, imagine that it's your responsibility to send a tremendous amount of data (say, an enormous database of historical statistics) from New York to London. The information must arrive as soon as possible, but once it's there, it won't require any additional updates or changes. If you're set on doing it electronically, the best solution might be one of the European T1 telephone lines that stretch beneath the Atlantic Ocean. With its data transfer rate of 2.048 megabits (Mb) per second, you would be able to transfer the contents of a compact disc (650 megabytes, or MB) in a little bit over 42 minutes.

The Concorde is roughly 6100 times faster and a 747 is 640,000 times faster than the electronic transfer

But if this database is truly enormous, the totally electronic solution can be beaten. A Concorde supersonic airliner can make the trip in 3 hours and 50 minutes carrying (along with passengers) a 1300-pound cargo payload, or about 32,500 CD-ROMs. This works out to a transfer rate of about 12.5 gigabits per second, roughly 6100 times faster than our electronic method. For a large data transfer job, a Boeing 747 freighter does even better. Although it takes more than seven hours to cross the Atlantic, the 747's higher payload of 248,300 pounds more than compensates for its slower speed. Its delivery rate of 1312 gigabits per second is more than 640,000 times faster than the wire. Slower but "wider" wins the race.

"Wider" Can Be Faster

	Transmission Method	*Transfer Rate*
	European T1 (1 CD-ROM)	2.048 Mbps
	Concorde (32,500 CD-ROMs)	12.54 Gbps (12,246 Mbps)
	Boeing 747 (6,207,500 CD-ROMs)	1,311.65 Gbps (1,280,913 Mbps)
	Container ship (2,240,000,000 CD-ROMs)	17,256.29 Gbps (16,651,852 Mbps)

A large container ship, "wider" still, takes eight days to cross the Atlantic, 27 times longer than a 747. But that ship may be able to hold 2.2 billion CD-ROMs—several years' worth of sales of audio CDs in the United States. Thus, the lumbering cargo ship wins the transatlantic data rate prize by delivering 17,256 gigabits per second—over 8.4 million times faster than the electronic method.

Carrying this to the extreme, the best method may be the oldest—a cargo ship

Our specific examples here are, of course, a little extreme.[1] For one thing, there may not be enough different CD-ROMs in existence to fill even a mere 747. For another, I've assumed that the Concorde, the 747, and the container ship were scheduled for your convenience, and I've left out the time it would take to load and unload them. Finally, there's the issue of routing: electronic communication is not limited by considerations such as whether the Concorde can land, how long it takes to get the data onto the discs, or how long it takes to move the discs from the container ship to some

The best solution would consider convenience, cost, and accessibility as well as bandwidth

1. Perhaps too extreme. This example assumes 25 CD-ROMs to the pound (why bother transporting the cases?); each disc contains 650 megabytes of data. For convenience, it uses published cargo weight capacities rather than volumes, which might be the actual limiting factor.

destination far from a seaport. But this example does show how you can trade speed of delivery for data capacity when looking for the maximum bandwidth.

The type of data being transferred may also influence selecting the best solution

The situation changes entirely if the value of the data depends upon its freshness. If the information will be useless after a few days or must be updated frequently, the cargo ship is useless. If the data must be updated interactively, a 28.8 Kbps modem would be much better than even the Concorde. As an extreme example, think about videoconferencing: quite a lot of data must be transferred to update the screen image each time one of the participants so much as smiles. And the fact that the value of data is often time-dependent is the reason that so much of business today is conducted via electronic means (whether by phone, fax, or computer) rather through the mail.

Still, in many situations, an overnight delivery service with a recordable CD-ROM or DAT (Digital Audio Tape) cartridge can outperform electronic delivery for speed, convenience, and cost. For instance, if you assume that a T1 connection to the Internet will cost $1,000 per month each for a person in New York and a colleague in London, then sending 100 GB of data (roughly the equivalent of 150 recordable CD-ROMs) would take about six days and cost $400 to transmit. Sending the same amount of data on recordable CD-ROM via Federal Express International Priority would cost about half as much and take only two days. And, if it simply must arrive as soon as possible, a courier on the Concorde will get there in about four hours but will cost over $5,000.[2] The following table compares the cost and the

2. This example uses somewhat different assumptions. The T1 (at a U.S. speed of 1.544 Mbps) in this example costs $1,000 per month to the Internet, and costs the sender and the recipient the same amount, thus $2,000 per month, or $66.67 per day, $2.78 per hour, and 4.6 cents per minute. A one-way ticket from New York to London on the Concorde costs about $5,400. The prices from FedEx are for priority (two days) and economy (seven days); this time the CDs are in jewel cases and weigh significantly more.

amount time required to send various amounts of data from New York to London via some common methods. The numbers are necessarily rough, but they give an idea of how the costs and times break down.

Comparing Costs and Times to Transfer Data

Amount of Data	*T1 (via Internet)*	*Concorde*	*FedEx Priority*	*FedEx Economy*
1 MB (1.4 MB floppy disk)	$0.004 (5 seconds)	$5,400 (4 hours)	$41.45 (2 days)	$28.50 (7 days)
100 MB (100 MB Zip disk)	$0.32 (7 minutes)	$5,400 (4 hours)	$41.45 (2 days)	$28.50 (7 days)
1 GB (2 650 MB CD-ROM discs)	$4.19 (91 minutes)	$5,400 (4 hours)	$41.45 (2 days)	$28.50 (7 days)
10 GB (16 650 MB CD-ROM discs)	$41.67 (15 hours)	$5,400 (4 hours)	$63 (2 days)	$59.75 (7 days)
100 GB (150 650 MB CD-ROM discs)	$419.44 (6.3 days)	$5,400 (4 hours)	$200 (2 days)	$183 (7 days)

For some years, the newsgroups of the Internet (Usenet) were distributed to Australia on data tapes sent from the United States by airmail rather than through a direct data link. This practice saved on transpacific telephone costs, and few users even realized that newsgroup postings weren't traveling via wires. Similarly, videotapes and films without immediate time value are still routinely sent to television stations and theaters by courier rather than through the more expensive and faster methods of satellite relay, microwave, or long-distance cable.

Even today, information is transferred on physical media

Keep in mind that bandwidth is only meaningful when it comes to sending information. You can't send actual goods, although you can send instructions for making them. Overnight

Bandwidth transmits information, not goods

delivery services may have been supplanted by fax machines and electronic mail for sending text documents, but when real objects are involved, traditional services—the post office, overnight delivery services, and even our friendly cargo ship—will always have a place. However, now that we have computer-controlled lathes and fabricating machines, it's hardly futuristic to imagine sending instructions for creating an object via electronic means.

Bandwidth Analogies

Bandwidth is like digital real estate

Bandwidth is often compared to real estate. Although the comparison isn't exact, there are many similarities, especially for bandwidth in the broadcast spectrum. Not only is the broadcast spectrum a fixed resource, its value depends greatly on location. Bandwidth in bands that are widely used—television, radio, cellular phone—is far more valuable than bandwidth in more obscure frequencies. And as with land, the federal government retains vast expanses of spectrum, especially for the military.

High frequencies are the vanishing frontier

There is even a vanishing frontier: high frequencies. As technology has made higher and higher frequencies useful, the amount of unused spectrum has shrunk to the vanishing point. So in some ways the higher frequencies—such as the microwave bands used for satellite broadcasts—resemble the frontier of the Old West. The assigned slots have wider bandwidth than slots in the older, lower frequency spectrum, just as the average real estate lot size in the western United States is larger than the lots in the east. The government gave large tracts of land to the pioneering railroad companies and has given away large amounts of spectrum to broadcasters. And like the railroads, some broadcasters are now trying to subdivide and sell portions of the spectrum they control.

Forms of Bandwidth

Bandwidth—in the sense of the means of transmitting information—comes in many different forms. The most obvious way to categorize them is by media:

- **Broadcast spectrum (over the air).** Bandwidth in the so-called radio spectrum is the most limited and tightly controlled, to prevent interference between two signals. The best-known uses of this spectrum are for one-way broadcast of radio and television either from terrestrial or satellite transmitters. But radio spectrum can also be used for two-way transmissions, such as cellular phones.
- **Telephone lines.** The ubiquitous telephone lines—pairs of copper wire—run from a telephone company switch (formerly called the "central office") to each telephone. Between the phone company switches, the wire is now either coaxial cable or fiber-optic cable, and it carries information in digital form. Coaxial cable can carry a few hundred simultaneous conversations; fiber is capable of carrying tens of thousands.
- **Cable television.** Cable television uses coaxial cable to carry multiple television channels to homes. The cable usually runs along telephone poles or underground, but every cable television subscriber in an area typically receives identical signals. That has begun to change as pay-per-view decoders and new digital set-top boxes allow individuals to choose what they will watch and Internet services begin to be distributed over cable television connections.
- **Local area networks.** In a local area network (LAN), coaxial, fiber-optic, or twisted-pair telephone cables interconnect an organization's computers so that information such as electronic mail and computer files can travel between machines.

- **Wide area networks.** Organizations with more than one location and groups of organizations that need to share information continuously use wide area networks (WANs) to span miles between sites or across continents. WANs typically use high-speed telephone lines to link the sites together. The Internet can be seen as a super WAN, a network of networks.
- **Packaged media.** A videocassette, a CD-ROM, a floppy disk, and a book all contain information. Although they may seem inert, you can think of them having bandwidth because there is always a rate at which information can be delivered from the package. For instance, an 8X CD-ROM reader might be able to deliver the information contained on the CD-ROM at a rate of 1200 kilobytes per second. A VHS player can deliver the cassette's information at a bandwidth lower than that of broadcast television.

Analog or Digital?

Analog data is represented in ways that correspond to its real-world existence

Information can be analog or digital. Analog information gets its name from working in ways analogous to the real world, where most of the things we experience vary in a continuum rather than in discrete jumps. In the real world, a louder sound consists of wider movement of vibrating air molecules in a sound wave. On an analog LP record, this is recorded as wider groove excursion, and on an analog cassette tape as a stronger magnetic field.

Digital data is represented by series of ones and zeroes

The digital world makes all its representations of the world with discrete binary digits (bits): zero and one, on and off. By combining and manipulating these bits, typically in 8-bit sequences called "bytes," computers are able to represent all manner of things. And because most computers work only with digital information, an ever-increasing amount of information now comes in digital form.

How and Why Analog Is Converted to Digital

The reason information from our analog world is often converted to digital form is simple: that is the only form that computers (and an increasing panoply of other devices, such as compact discs and DVD players) can understand. To convert an analog signal—say, an audio waveform—an analog-to-digital **sampling** circuit takes rapid samples ("snapshots") of the analog signal. Each sample is converted to a number that is stored in digital form. But unlike a continuous analog signal, a digital signal is limited to a finite number of discrete steps that relate to the amount of storage it takes up.

Sampling is used to convert analog data to digital form

That finite number depends on the system. For example, in an 8-bit digital audio recording, the loudest sound might be sampled as a 11111111, the softest as 00000000, with just 254 discrete steps in between. A 16-bit system using the same technique would offer 65,536 steps and a theoretically smoother sound. In both cases, the sampling would occur thousands of times a second to capture the analog sound. And in both cases, the process would simply be reversed to recreate the original analog signal for reproduction.

A 16-bit system theoretically produces a smoother sound than an 8-bit system

Figure 3-1 *Analog to digital sampling and conversion.*

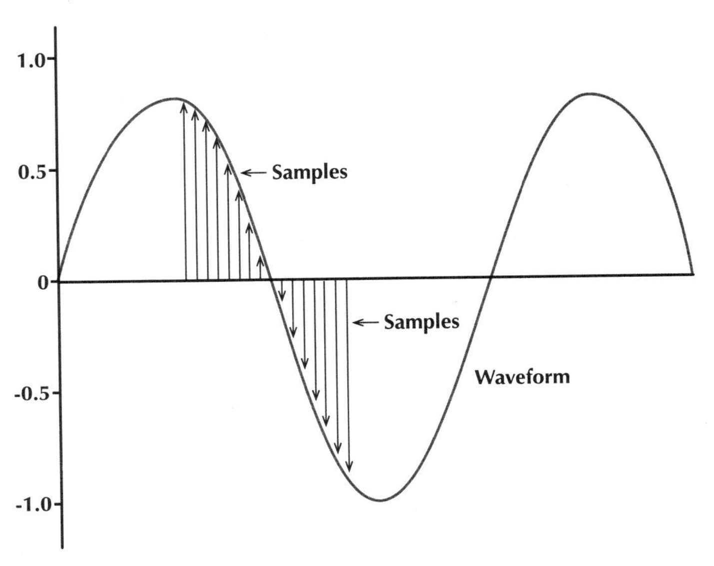

Digital Information Is Not Better Than Analog

Digital data is not inherently better than analog

Despite the hype, information in digital form is not inherently better than the same information in analog form. In fact, an argument could be made that an analog original always contains more information than its digital equivalent, because the sampling process by definition breaks an analog signal into pieces, and you can always imagine smaller pieces. But analog systems, like vinyl records, are often plagued by noise that digital systems can avoid.

A noticeable difference in quality is often a matter of bandwidth

When people judge digital versus analog and declare one kind of information better, they're generally comparing high and low bandwidth examples. Compact discs, with their high-bandwidth digital audio signals, sound better than analog cassette tapes. But a digital cellular phone call is typically poorer in quality than an analog cellular call, because the digital signal uses only half the bandwidth of the analog call. Similarly, digital cameras create lower-quality images than traditional analog film cameras. This is because the resolution of film contains more information than its digital counterpart, so the analog film has greater bandwidth and hence better image quality. Digital cameras, particularly very expensive ones, are improving, but analog camera images are likely to remain superior for many years to come.

The Advantages of Digital

Digital signals are more resistant to distortion and interference

Digital signals do have several advantages apart from the fact that computers can process them. Digital information is easier to store, manipulate, and transmit. In order for data to be transmitted, it must be duplicated. Analog signals cannot be duplicated perfectly because the copying process always adds at least a little distortion. Digital signals are much more resistant to distortion and can usually be copied perfectly. And digital signals can be intermixed easily, so your disk drive or the Internet can carry electronic mail, Web pages, voice conversations, and video all at the same time. Analog signals usually require a specific circuit or recording device for each type of signal.

There is another digital advantage: a high-bandwidth analog signal, such as a television program, is virtually impossible to transmit over a low-bandwidth connection, such as a radio station or a telephone line. However, once information has been converted to digital form, it can travel on any digital link. A full-motion video signal can travel over a low-bandwidth modem connection, albeit slowly. A one-minute video might take an hour to send, but once it arrives, it should look just as good as if it were sent over a high-bandwidth line.

Digital data can travel across any digital link

But waiting an hour for a minute of video isn't high on anyone's list of fun ways to pass time. From our experiences with telephones, television, and radio, we've become accustomed to rely on "real-time" connections, in which that minute of video would take precisely a minute to arrive from the sender. A live audio or video transmission is a specific case of a real-time connection that requires a high-bandwidth line. One goal of the so-called information superhighway (a term that has generally passed out of favor over the last few years) is to have enough bandwidth so that everything, including full-motion video, can be handled in real time. We are still a long way from this goal today, but much of the development in personal computers and advanced networking technologies over the next few years will lay the groundwork for real-time audio and video in the future.

Real-time transmission of live data requires high bandwidth

One More Digital Advantage

There is one more advantage to a digital signal, and it has enormous ramifications for bandwidth today. Analog signals are highly susceptible to interference from one another, as anyone who has tried to tune in a distant radio or TV station knows well. For that reason, analog signals are strongly separated from one another as much as possible, as in the electromagnetic "bands" that gave bandwidth its name. It is

Analog signals are separated to reduce interference

extraordinarily difficult, for example, to mix two analog conversations in one portion of the spectrum and somehow separate them at the other end.

Digital signals can be intermixed and decoded

But because they are merely ones and zeroes, digital signals can be mixed together and transmitted in a variety of ways, as long as some additional information is provided so that the signals can be put together properly at the receiving end. The implication of this is that many separate messages may travel at virtually the same instant over the same medium and arrive at a single destination (or many separate destinations), where they can be decoded properly. Understanding this sharing of data is essential to understanding the way computer networks, and most particularly the Internet, work.

The phone network is known as "circuit-switched" because your call essentially occupies one complete circuit from start to finish. Over a standard line, your call effectively "owns" a line dedicated to it (although in reality it's a little more complicated), so nobody else's information can intrude upon it.

Millions of separate packets travel simultaneously without interfering with one another

But digital communications are typically "packet-switched." At the transmitting end, the information is broken up into tiny "packets" that contain not only the data you want to send, but also information about the packets themselves, where they should go, and how to put them together again at the receiving end. Once they are out on the network, it's the network's job to route them in the proper direction. But the key point is that the packets you're sending are traveling along with millions of others. In this scheme, you no longer control the data pipe for a limited amount of time; instead, you are sharing a much bigger pipe with millions of others.

This can be highly efficient, because when your data is not flowing through the system, others can use it to send their information. But it can also create problems. If everybody wants to download Mars pictures at once, the data glut may slow everyone down to a crawl, and you may have to get in line. In addition, the information about where to send your packets and how to reassemble them wastes some of the raw bandwidth. Despite these problems, packet-switched networks—including the Internet—are clearly the wave of the future.

Sharing bandwidth is efficient even though a little more space is used for the extra identifying data

How Packets Work

A **packet** is the smallest unit of data that makes up all communications on the Internet. For instance, when you send an e-mail message, it's broken into packets, and then each packet is sent separately to the eventual destination, where all the packets are reassembled into the original e-mail message.

Although the details vary among network protocols, every packet consists of at least two parts: the data and the header. The header of a packet includes information needed for transmission and reassembly at the receiving end, information such as the length of the packet, the destination address, and the source address. Network devices look at the destination and source addresses and route the packet appropriately. Although all the packets that make up an e-mail message, say, are likely to take the same route through the Internet, that's not necessary, and the separate packets can be routed in different ways, as long as they all arrive at the destination for reassembly into the whole.

In Conclusion

This chapter looked briefly at the many different ways to think about bandwidth. It's a subtle topic, and one that many people fail to understand because of the misleading terms like "speed" and "faster" that are bandied about with no regard for reality. Speed is always a measure of distance divided by time, whereas bandwidth is a measure of information flow over time.

Now that you have the right mindset, let's delve more deeply into the details of how bandwidth works and the terminology used to describe it.

Chapter Four

Looking at Bandwidth

This chapter discusses the basic issues of bandwidth. Some of the material is a little technical, but it should help you understand the limitations of different methods of communication. If you have some familiarity with analog and digital hardware and reading the specifications for a stereo system, you may want to skip to the next chapter and come back here for reference, if necessary.

Analog Bandwidth

Analog bandwidth is usually defined as a frequency range in terms of **hertz**, or Hz, a unit named for Heinrich Hertz, who was the first to create and detect radio waves. Hz, appropriately, is a unit that measures waves in terms of cycles per second (and used to be called, straightforwardly enough, "cycles per second"). A cycle is measured from crest to crest or trough to trough; how many of them pass a fixed point in a second, whether hundreds, thousands, or billions, is the number of Hz involved. All else being equal, the higher the Hz, the shorter the wavelength, as shown in Figure 4-1.

Analog bandwidth is a frequency range measured in cycles per second

Figure 4-1 ***Hertz is a unit that measures waves in terms of cycles per second.***

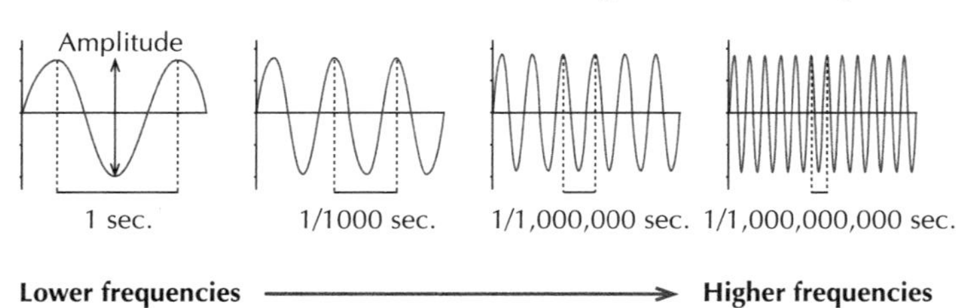

A telephone call delivers sound roughly between 200 Hz and 3200 Hz. The range between 200 and 3200 is 3000, so a telephone call has bandwidth of 3000 Hz or 3 kHz. This figure is approximate; the exact number depends on the phone system and the particular components of the telephone in question, including the mouthpiece and the earphone. Pianos are tuned such that middle A resonates at 440 Hz. The table on the following page shows the frequency ranges of several common instruments.

Audio Quality and Analog Bandwidth

More bandwidth can mean better analog audio quality

A telephone cannot deliver high-fidelity sound; 3 kHz simply isn't enough bandwidth, and a telephone signal suffers from a host of distortions in addition to limited bandwidth. An AM radio broadcast sounds better than a telephone call because it has about 5 kHz bandwidth. An FM radio signal, with 15 kHz bandwidth, sounds even better and meets the minimum standards for high fidelity.

Radios extract the audio signal from the carrier frequency

For radio broadcasts, these bandwidths are superimposed on a **carrier frequency**. When you tune an AM radio station to a carrier frequency of 710 kHz, for example, the receiver actually pulls in a signal that the station is transmitting from 705 to 715 kHz. Your radio detects the carrier frequency and extracts the difference between the carrier frequency

The Frequency Range of Common Instruments

Instrument	*Note Range*	*Typical Fundamental Frequency Range*
Piano (88 key)	A(0) to C(8)	27.5 to 4000 Hz
Harp	B(0) to G#(7)	30.87 to 3322 Hz
Pipe Organ Pedals (16' pitch)	C(1) to G(4)	32.7 to 392 Hz
Tuba	D#(1) to F(4)	34.65 to 349.2 Hz
Double Bass	E(1) to D(4)	41.2 to 293.7 Hz
French Horn	Bb(1) to F(5)	58.27 to 698.5 Hz
Tympani	D#(2) to B(3)	77.78 to 246.9 Hz
Electric Guitar	E(2) to E(6)	82.41 to 1318.5 Hz
Marimba	C(3) to C(7)	130.8 to 2093 Hz
Clarinet	D(3) to C(7)	146.8 to 2093 Hz
Trumpet	E(3) to D#(6)	164.8 to 1244.5 Hz
Violin	G(3) to F(7)	196 to 2794 Hz
Flute	C(4) to C(7)	261.6 to 2093 Hz
Pipe Organ Manuals (8' Pitch)	C(4) to C(7)	261.6 to 2093 Hz

and the signal to recover a 5 kHz audio signal. (The section "AM Radio," in the next chapter, explains why the broadcast signal is 10 kHz wide instead of 5 kHz.)

Audio Quality and Dynamic Range

A compact disc offers bandwidth of about 20 kHz. It does not have significantly higher bandwidth than FM radio, primarily because only young people can hear much above 15 kHz. But the compact disc can sound better, and its main improvement in audio quality is **dynamic range**, the range from the loudest to the softest sounds the medium can produce. The dynamic range of a signal measures the difference between the highest possible peak signal and the lowest detectable signal; the wider the range, the better.

The dynamic range and signal-to-noise ratio also affect audio quality

A closely related concept is the **signal-to-noise ratio** (SNR). The SNR measures the ratio of the desired signal (music, voice, picture) to the noise that all communications channels have. You can hear this noise when you listen to a distant AM station or when you are talking on a cellular phone at the edges of the coverage area, and you can see it when looking at a snowy picture on a television set. The SNR measures noise relative to a signal level that usually isn't a peak level. With both dynamic range and SNR, the higher the measurement, the better the quality. Some people consider the two terms to be functionally equivalent; in this book, however, the distinction will be maintained.

Small improvements in SNR can improve sound quality considerably

Dynamic range and SNR are both measured in **decibels** (dB), which means one-tenth (deci) of a bel, named for Alexander Graham Bell. The decibel measurement is logarithmic, so a 10-dB improvement is much more than 10 times better than a 1-dB improvement. A 10-dB improvement in the SNR is very appreciable, at least up to 70 or 80 dB. For example, Dolby B noise reduction for cassette tapes improves the SNR only 5 to 10 dB, yet it improves the sound quality considerably.

Improvement past 80 dB is usually difficult to hear because the inherent noise of the medium is lower than the ambient noise at either the source or the listening room. If you hear noise on a digitally recorded CD, it may well be from an air conditioner in the recording studio or traffic outside it.

Bandwidth and dynamic range usually go together

The examples in the following table will give you a sense of the dynamic range of different media and how they relate to the medium's bandwidth. As you can see in the examples, bandwidth and dynamic range are generally directly related to one another, and an improvement in either one (or the SNR) can improve quality.

The Bandwidths and Dynamic Ranges of Common Audio Media

	Bandwidth	*Dynamic Range*
Telephone call	3 kHz	30 dB
AM radio	5 kHz	40 dB
FM radio, cassette	15 kHz	55 dB
Hi-fi audio tracks, VHS cassette	18 kHz	80 dB
Audio CD	20 kHz	96 dB

Exceptions to the Rule

To provide the best quality, an analog medium should have both wide bandwidth and wide dynamic range. Although the two criteria usually go together, there are some exceptions. A radar or sonar sending and receiving system needs only a very narrow bandwidth—just wide enough to monitor the ping that the system sends out and the returning echoes—with wide dynamic range to detect very faint echoes. One example of a very wide bandwidth combined with very low SNR comes from astronomy. The spectral analysis of very distant stars consists of examining a wide bandwidth signal with a very high noise level; the spectra must be sampled for a long time to extract information from the abundant noise.

In some cases, wide bandwidth, wide dynamic range, or very high SNR alone can yield a high-quality signal

Video Bandwidth Measurements

Video bandwidth is a more complex topic than audio bandwidth because there are three common ways to quote analog video bandwidth: the bandwidth of the channel, the bandwidth of the video signal, and the number of lines of resolution. The first describes the bandwidth of a television broadcast channel, which is 6 MHz for the National Television System Committee (**NTSC**) television system that is used in North America. The 6 MHz includes the sound, the video, and some margin to separate the channels. This method of defining analog video bandwidth works only for broadcast video, not for recorded video such as is found on a VHS

Analog video bandwidth may refer to the bandwidth of the television broadcast channel, the bandwidth of the video signal alone, or the number of lines of screen resolution videotape or a laserdisc. A second way to define video bandwidth is by the video signal alone, which has a bandwidth of 4.2 MHz. In the marketplace, the common specification for video bandwidth is the number of lines of resolution along a horizontal scan line, and this is the third way of quoting analog video bandwidth.

Higher bandwidth does not always mean a higher SNR

As with audio signals, the signal-to-noise ratio also affects quality, but unlike audio, SNR doesn't correspond exactly with bandwidth. The Bandwidth Measurements and SNR for Common Video Sources table compares different measurements of bandwidth (not including the channel bandwidth, which is only relevant for broadcast video) and SNR for three common analog video sources. As you can see, the correspondence is not exact. An increase in bandwidth does not necessarily mean that the SNR will increase.

Bandwidth Measurements and SNR for Common Video Sources

	Video Signal Bandwidth	*Horizontal Lines*	*SNR*
VHS videotape	2 MHz	230 lines	43 dB
Broadcast TV	4.5 MHz	330 lines	50 dB
Laserdisc	5 MHz	400 lines	47 dB

Although SNR does not track precisely with bandwidth as measured either by video signal or horizontal lines (largely because different systems introduce their own forms of noise into the signal), it's clear that quality is related. The picture quality of broadcast television is better than that of VHS videotape, and the picture quality of laserdisc is better yet. The wider the signal bandwidth or the more lines of horizontal resolution, the higher the picture quality.

Digital Bandwidth

Unlike analog bandwidth, which is a measurement of the amount of spectrum each signal occupies, digital bandwidth is a measurement of the amount of information each signal carries. Digital bandwidth is specified in bits per second. The more bits per second, the more information. Each bit represents a zero or a one (off and on) in binary numbers. Because digital signals have lots of bits, the common way to measure digital bandwidth is in kilobits (thousands of bits) per second (Kbps), or megabits (millions of bits) per second (Mbps). Some very high bandwidth communication is measured in gigabits (billions of bits) per second (Gbps). See the Measurements of Digital Bandwidth table for some common measurements.

Digital bandwidth is measured in kilobits per second (Kbps) or megabits per second (Mbps)

Measurements of Digital Bandwidth

Bandwidth Measurement	*Exact Measurement*	*Vernacular Measurement*
bps (bits per second)	1 bps	1 bps
Bps (bytes per second)	8 bps	1 Bps
Kbps (kilobits per second)	1024 bps	about 1 thousand bps
KBps (kilobytes per second)	8192 bps	about 8 thousand bps
Mbps (megabits per second)	1,048,576 bps	about 1 million bps
MBps (megabytes per second)	8,388,608 bps	about 8 million bps
Gbps (gigabits per second)	1,073,741,824 bps	about 1 billion bps

In principle, the digital bandwidth—the bits per second—can fully describe the capabilities of a digital signal. The single number ought to tell us whether the system can handle a particular kind of digital signal. But because bandwidth can be expressed in several ways, it is sometimes quite misleading. Throughout this chapter you will find that some specific ways measurements of digital bandwidth can be confusing.

Digital bandwidth measurements can be misleading

Bits vs. Bytes

Eight bits equal one byte

One bit cannot convey much information, so a computer usually deals with a group of bits at a time. A single character (a letter of the alphabet, a number, or a punctuation mark) is coded by eight bits in sequence, or one byte. The letter A, for example, is coded as 01000001, B is 01000010, and so on. Most computing devices store and transmit data in this code, which is called **ASCII** (American Standard Code for Information Interchange).

Multipliers are misleading: 1 kilobyte = 1024 bytes, not 1000 bytes; and 1 megabyte = 1,048,576 bytes, not 1 million bytes

The information content of a single byte is still limited, so information is often measured in kilobytes, megabytes, or gigabytes. There's some confusion about the multipliers. Although *kilo* ordinarily means 1000, a kilobyte is not an even 1000 bytes because the computer's counting system is based on the number 2, not the number 10. The power of 2 nearest 1000 is 2 multiplied by itself 10 times (2^{10}), or 1024. Although technical literature generally uses 1024 as the multiplier for kilo, casual writing often uses the simpler 1000. Similarly, *mega* would be 1024 times 1024 or 1,048,576 rather than exactly one million. These discrepancies aren't usually serious, but they do explain why the numbers used to describe computers and digital communications often don't quite match up, as you can see in the Digital Storage table on the following page.

Common practice specifies data communication speed in bits per second, and digital storage capacity in bytes

Even more confusing is the fact that although data communication is usually measured in bits per second, other measurements of digital bandwidth commonly use bytes, so when you see a number for digital bandwidth, you must check to see whether it's in bits or bytes. Often, the sales person quoting these figures won't know whether they are in bits or bytes. This book follows the common practice of specifying data communications speed in bits per second, and digital storage capacity in bytes, making sure to specify bits or bytes precisely in cases where there might be confusion. In abbreviations, bits are lowercase b, and bytes are uppercase B. Thus bits per second is bps, bytes per second is Bps. Remember, eight bits is one byte.

Measurements of Digital Storage

Storage Measurement	*Exact Measurement*	*Vernacular Measurement*
b (bit)	1 bit	1 bit
B (byte)	8 bits	1 byte
KB, or K (kilobytes)	1024 bytes	about 1 thousand bytes
MB (megabytes)	1024 kilobytes or 1,048,576 bytes	about 1 thousand kilobytes or 1 million bytes
GB (gigabytes)	1024 megabytes, 1,048,576 kilobytes, or 1,073,741,824 bytes	about 1 thousand megabytes, 1 million kilobytes, or 1 billion bytes (In British usage, a gigabyte is a milliard bytes.)
TB (terabytes)	1024 gigabytes, 1,048,576 megabytes, 1,073,741,824 kilobytes, or 1,099,511,627,776 bytes	about 1 thousand gigabytes, 1 million megabytes, 1 billion kilobytes, or 1 trillion bytes (In British usage, a terabyte is 1 billion bytes.)

Bits Per Second and Baud

Baud is a nineteenth-century telegraphy term describing data transmission speed that was used in describing the bandwidth of early modems. Strictly speaking, one baud is one signal event or modulation change per second (a *symbol* in communication theory). For a 300-baud modem, one baud—one signal event—corresponds to 1 bit per second; at higher speeds, one signal event encodes 2, 4, or more bits. A 2400-bps modem actually runs at 600 baud because it sends 600 events per second; each event encodes 4 bits. Many writers don't understand the distinction and use baud interchangeably with bits per second, even though baud is essentially an incorrect term at all speeds above 300 bps. The term *baud* comes from J.M. Emile Baudot (1845–1903), who developed a five-bit code—the Baudot code—for the alphabet, a precursor to the ASCII code.

Pictures take up more storage space than plain text

A double-spaced typewritten page contains about 1500 characters, which takes up 1.5 kilobytes or 12,000 bits if the data is simply text. However, a fax of the same page takes up much more storage space, roughly 240 kilobytes. Why? Because a fax is a picture of a page, and a picture of the letter E requires many more bytes than the ASCII code for E, which requires merely one byte (see Figure 4-2). On the other hand, the ASCII code contains no information about the font and size, while the fax shows what each letter looks like, albeit rather coarsely.

Figure 4-2 ***The ASCII code (left) for the letter E contains 8 bits, while the fax image (right) of the letter E contains 256 bits.***

01000101

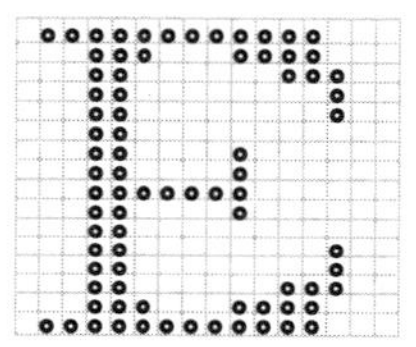

So, varying amounts of space are required for handling different types of data. The What a Single Megabyte Can Store table gives you some idea as to what can be contained in a single megabyte without the use of compression.

What a Single Megabyte Can Store

Books	One long novel, stored as text
Pictures	One full-page black-and-white image One 3-by-5 inch color picture
Sounds	Two minutes of telephone–quality sound Seven seconds of compact–disc quality sound
Video	0.04 seconds of broadcast–quality video

How Digital Signals Are Carried Over Analog Channels

Digital signals can be carried over analog communications channels. A modem converts digital signals to sounds so that they can travel over an analog telephone line; the name **modem** means modulator/demodulator. A modem modulates an analog signal with digital information when sending and demodulates the analog signal to recover the digital information when receiving. Digital signals can also be embedded in radio and television broadcasts or transmitted via cable television cables.

Comparing Analog and Digital Bandwidth

There's no simple way to compare digital and analog bandwidths for a communications channel. In early modems—up to 300 bps—the digital signal was sent as a series of tones over the phone line (reversing the original process by which the telegraph led to the telephone). When coded as tones, the analog bandwidth or frequency range directly limited the digital bandwidth. But in all recent modem standards, the digital signals use not only frequency but also a variety of other tricks to encode data, such as changing the amplitude and the phase of the signal. In short, there is no longer a direct connection with the traditional analog bandwidth and digital communication speed over that channel.

When coded as tones, digital bandwidth over phone lines was limited

Converting Analog Signals to Digital By Sampling

Channels designed specifically for digital communications generally do not have the characteristics necessary for carrying high-quality analog signals. Instead, the analog signal must first be digitized for transmission. **Digitizing**, also known as **sampling**, is the process of converting an analog signal to digital form by recording the state of the analog waveform (taking a sample) at regular and frequent periods. Some types of communications have switched from analog to digital without most users' realizing it. Most telephone calls are

Many communications have switched from analog to digital without our noticing analog only between the telephone company central office (switch) and the telephone; this part of the connection is called the local loop. Everywhere else, the telephone call is carried in digital form. Only local calls placed on older telephone company switches remain purely in the analog domain.

The higher the sampling rate and depth, the better the quality and the higher the cost

The quality of conversion from analog to digital is determined by the digital **sampling rate** and **sampling depth**. The sampling rate measures how frequently the analog waveform is converted to digital form. To capture an analog waveform, the sampling must be at least at twice the highest frequency of interest. So, to capture a 20 kHz tone, the sampling must be made at 40 kHz; each wave must have at least two data points. The sampling depth is the number of bits that characterize a single sample; the greater the bit depth, the greater the signal-to-noise ratio. The more bits, the more accurately the waveform can be characterized. The higher the sampling rate and the greater the bit depth, the higher the quality—and the higher the costs, with more bits to store and distribute.

The sampling rate for CD quality audio is 44.1 kHz

For compact disc audio, the minimum standard was set for an analog bandwidth of 20,000 Hz and an SNR of 90 dB. This led to a sampling rate of 44.1 kHz (a little higher than 40 kHz, to give the electronics some margin to work with when converting back to analog audio), and a sampling bit depth of 16 bits or 65,536 levels, which corresponds to a 96-dB dynamic range. A sampling rate of 44.1 kHz, multiplied by 16 bits, multiplied by two channels for stereo, yields 1.41 Mbps; with some overhead, the full bit rate is 1.5 Mbps.

Lower sampling rates for audio include 768 Kbps for FM stereo quality and 56 Kbps for telephone quality sound

CD-quality sound is not always necessary, so lower sampling rates with shallower bit depths suffice for many applications. For FM radio quality sound, sampling is often set to 32 kHz with a 12-bit depth, producing a 384-Kbps data rate (in mono—double this for stereo for a data rate of 768 Kbps). Telephone calls are normally digitized at 56 Kbps and are usable even at a data rate as low as 16 Kbps.

Problems with Analog Bandwidth

Communications channels are never perfect. Much of the engineering effort goes into coping with problems that intrude into the signal. This discussion deals with the problems that arise with analog signals, even when everything is working correctly. Noise, distortion, interference, breakdown, interruptions, crosstalk, wow, and flutter are among the problems you can encounter.

Noise There are many sources of intrusive noise when something fails in an analog signal, such as poor contacts in plugs, faulty components, moisture in the wires, or animals gnawing on cable. This noise can render an audio signal unusable. These conditions can also afflict digital channels.

Noise is inherent in all analog systems. At any temperature above absolute zero, there is thermal noise in electronic equipment. When capturing images from faint stars, astronomers use cooling devices to reduce thermal noise in solid state charge coupled device (CCD) cameras. For less exotic equipment used at room temperature, the noise can be quite low, but it never disappears. Each time the signal passes through another circuit, new noise is introduced.

Noise is everywhere, all the time

Every time an analog signal is copied, the noise level goes up. Every time a telephone call goes through a relay amplifier, the background hiss goes up a notch. When the phone system was all analog (through the 1980s), you could tell if a call was long distance by the hiss in the background; a long-distance call went through more amplifiers than a local call, and therefore had more hiss. Today nearly all long-distance calls are relayed in digital form, so the hiss has disappeared.

The amount of noise in an analog signal increases with every copy made

Whenever a tape or film is copied (dubbed), the noise level goes up—producing more hiss in tapes, more grain in film. The original professional Dolby noise reduction system (*Type A*) for recording studios was especially important because it permitted several successive generations of tape copies for editing and processing the sound. In addition to the increase in

Tape and film dubs have more noise, grain, and contrast

background noise, the dynamic range is altered a little with every copy. For photographic copies, the contrast goes up; that's why old movies usually have such high contrast. Modern motion pictures are printed with special low-contrast printing masters designed to minimize the contrast buildup.

Irregularities, dust, scratches, and print-through all cause signal interference in analog storage media

Interference and breakdown Analog storage media all have problems that interfere with the signal. Irregularities and dust on magnetic tape can produce output variations and dropouts, where the signal simply disappears for a moment. Over time, one layer of magnetic tape can print through to the next, producing a faint echo or pre-echo, depending on which layer prints through. Physical scratches on videotape can show up as lines on the screen. Pops, ticks, and scratches mar the music stored on a long-playing vinyl record. Dust and scratches are constant problems for photographic film. In a movie theater, you can tell when the film reel is about to end when a blizzard of color confetti appears; the ends of a reel are exposed to more dirt and scratches than the middle of the reel.

Distortion levels depend on the signal levels

Distortion Distortion is also inherent in analog systems. No electronic circuit can preserve a signal exactly, although for many circuits the added distortion is not perceptible. But if an analog signal passes through hundreds of circuits before reaching its destination, the distortion can build up; a single bad circuit will damage the signal irreparably. In all devices, the level of distortion depends on the signal level. Very loud signals distort far more than quiet signals, which explains the distortion you hear when you turn up the volume too high.

Distortion in stereo equipment is indicated by a percentage, but some distortions are more audible than others

The most common measurement of distortion is the **harmonic distortion** figure quoted for stereo equipment, which refers to the presence of frequencies in an output signal that were not present in the input. The amplifier in a typical modern stereo receiver might have 0.1 percent distortion over the bandwidth of 20 to 20,000 Hz. A cassette tape recorder has much higher distortion, often 3 percent or more for loud signals. Good speakers typically exhibit 10 percent distortion. A cheap boom box will easily produce 100 percent or more distortion; the distortion is stronger than the original signal, which will still be recognizable.

The percentages don't tell the whole story; some distortion is much more audible than others. Psychoacoustical testing shows that in careful listening, people can hear 0.7 percent distortion with music and 0.3 percent with pure tones that no one ordinarily listens to. Thus tape recorders, speakers, and microphones routinely have distortion levels well above this threshold. The control and shaping of distortion into acceptable form is what analog equipment design is all about.

In some situations, people seek to add distortion to a signal. Many musicians use electronic guitars with accessories—*fuzz* and *grunge* effects and the like—that introduce deliberate distortion. Some devices boast that they produce "the most brutal distortion ever made." These distortion effects are shaped, and more or less controlled, with the music, in a manner that's usually different from ordinary distortion. Indeed, the reason a violin and an oboe sound different is that each instrument produces a distinctive distortion overlaid on the basic note.

Some distortion is deliberate

Response curves of equipment How well does a circuit or communications channel respond over the analog bandwidth? In stereo equipment, **flat response** from 20 to 20,000 Hz has always been the goal for amplifiers. This means that all signals over the bandwidth are amplified equally. In cheaper amplifiers, the response might be far from flat. Speakers are never flat; what's more, speakers interact with the room they are in and so their response curve depends on their placement, particularly at the lower frequencies.

Stereo amplifiers are engineered for flat response, meaning all signals are amplified equally

When frequency response is not flat, the error is indicated by the amount of decibel variation

The response curve over a given bandwidth is often specified with a decibel variation, such as *20 to 20,000 Hz plus or minus 1 dB*. If no plus or minus decibel figure is given, the device most likely has poor performance. A company might claim that a particular speaker reproduces sound from 50 Hz to 15,000 Hz, a claim that may hide serious irregularities within that bandwidth, and the output at 50 Hz may be 20 dB down from the output at 1000 Hz. A signal that is 20 dB down from the peak is almost inaudible. A fair specification for such a speaker might be 120 Hz to 12,000 Hz, plus or minus 5 dB. No speaker is flat to within 1 or 2 dB, unlike stereo amplifiers, which are typically quite flat.

Analog signal interruptions may be impossible to repair

Interruptions An analog signal cannot be interrupted without creating difficulty. If a radio program is stopped halfway and resumed later, the only way to restore the complete program is to edit or splice the parts together—an inconvenient and difficult process. If the break occurs in mid-sentence, even a skilled editor will have trouble. Radio studios obviously have the tools to do this, but very few radio listeners can, or want, to edit programs. Generally, because an analog signal is normally **synchronous**, an interruption in an analog signal causes a delay.

Synchronous means that whatever you send must be sent in real time; a half-hour television show is sent in half an hour, and demands analog bandwidth able to handle a television signal. There are some exceptions to this; a signal can be taped and played back at slow or fast speed. Double cassette decks usually have a 2X dubbing function that copies a tape at twice the normal speed.

Crosstalk and interference refer to signal leakage through a wire and through the air, respectively

Crosstalk and interference These occur when some other signal leaks into the desired signal. There's no clear distinction in the terms, although crosstalk usually means signal leakage through a wire and interference usually means leakage through the air, such as interference between two radio stations. Some kinds of crosstalk are not very serious: crosstalk between the left and right channels of a stereo signal, for example, reduces the stereo effect, but few listeners will notice. The worst interference takes place in the international short-wave radio bands, where stations choose frequencies and schedules on their own without any overseeing authority.

Wow and flutter These are pitch distortions that result from uneven speeds during the recording and playing back of magnetic tape or phonograph records. Wow typically describes lower-pitched variations; flutter describes higher-pitched ones. The sustained tones in piano recordings are particularly sensitive to wow and flutter: in severe cases they produce an underwater-like effect. There is no way to correct this problem for audio signals.

Wow and flutter are the result of speed irregularities; audio signals cannot be corrected, but a time base corrector can fix the problem for video signals

On a VCR, similar speed irregularities produce an unstable picture that jumps or drifts. But unlike audio recordings, a video signal has clearly defined time markers for each frame and line, making correction possible. A device called a time base corrector can fix the video problem by replacing all the synchronizing signals and retiming the picture information. Nearly all professional video systems include time base correctors, but very few consumer video products do.

Benefits of Analog

This long laundry list of problems with analog signals describes only problems that are inevitable when everything is working correctly. When you consider all the things that can go wrong, it's a tribute to engineers that analog signals work at all, much less survive being sent around the world or from the moon to the earth. It's equally impressive that something as inherently clumsy as a phonograph record can not only reproduce beautiful music, but do so well enough that some people actually prefer LP records to compact discs, believing that LPs retain information that is often lost in the interstices of the digital sampling process.

Film cameras capture images with better resolution and tone than do digital cameras, and at a lower cost

Some types of analog information storage will persist long into the digital era. Photographic film captures images with a resolution and tonal rendition at a cost that no digital camera will come close to matching for decades. Nevertheless, digital cameras are already popular because they produce pictures quickly, and many pictures, particularly small ones displayed on video screens, do not require the image quality of photographic film.

Benefits of Converting Analog to Digital

Early conversion from analog to digital minimizes the accumulation of analog problems

Once in the digital domain, a signal is handled very differently from analog. Most audio and video signals start out first in analog form, and that signal may not be free of analog problems. However, if the signal starts as analog, immediate conversion to digital form will prevent many analog problems from accumulating. Some signals start in the digital domain—for example, information generated in a computer, computer-synthesized music, and computer animation. On the receiving end, the digital signal is converted back to analog for us to hear or look at, again with an analog path that is short and generally easy to manage. The goal in maintaining quality is to keep the analog portions to a minimum at both ends.

Sampling rate and depth are critical

The conversion from analog to digital and back again is a critical step; the sampling rate and depth determine the quality of the signal that can be recovered. As explained earlier, this choice is a compromise between quality and data rate, where high quality means many more bits per second, demanding higher, and therefore more expensive, digital bandwidth.

Problems with Digital Bandwidth

Once a signal is digitized, the many analog problems evaporate, replaced by new ones unique to the digital world. Some problems are in the technology, somewhat like analog noise and distortion problems. Error correction and digital noise fall into this category. Other problems relate to the effects of compression, costs, and politics.

Noise and error correction in digital signals Digital communications channels operate with a very small signal-to-noise ratio (SNR) if measured by analog standards. On a 28.8 Kbps modem connection, the effective signal-to-noise ratio between the data and noise is minimal—only two or three decibels. For digital data, the SNR doesn't affect the quality of the signal, as long as the original data can be recovered. However, for a fixed-capacity channel such as a telephone line, the highest bandwidth can be achieved with the smallest possible SNR.

Digital transmissions can tolerate low SNR

Digital systems are usually able to correct for errors in communication. The standard method has the originating computer calculate a number based on the digital information (the **checksum**) and then send the information along with the calculated number. The receiving computer takes the information, makes the same checksum calculation independently, and compares its results with the sending computer's checksum. In case of a discrepancy, the receiving computer asks for the information to be sent again.

Checksums are used to detect digital communication errors

This interactive method is used by many error correction protocols, including early ones such as Xmodem and Zmodem. But these early interactive error-checking methods were inefficient, because both the request for resending and the repeated information itself had to traverse the entire data path. Most digital communication involves several different steps, and the problems, if any, typically occur at only one or two steps. Thus, in modern communications, error checking occurs independently at each stage, and any problems are resolved at those stages, without involving the entire path.

Modern error correction protocols check for trouble at each stage

The computers that we think of as the sending and receiving units may not be involved at all in the error correction. Whereas early interactive error-checking methods required a computer to do the work of verifying the data, modern modems on either end of the connection do the error checking and retransmission, if necessary. This speeds the process by eliminating the need for a computer-to-modem communication on either end when retransmissions are necessary.

Error correction reduces throughput

Any error correction information takes time, and therefore bandwidth, reducing the overall **throughput** of the system. Invoking error correction can be avoided if the data transmission rate is slowed to make the signals clearer. The trick is to send data in such a way that the overall throughput is the highest possible. In recent modem protocols such as V.32bis and V.34, the modems test for line quality and adjust their speeds up and down for maximum throughput.

Forward error correction involves sending redundant information in case a problem occurs

Interactive error checking obviously works only when the receiving device can request the sender to repeat information. If the communications channel is not interactive, error correction is done by sending redundant information in advance, so that the receiver can calculate the correct data even if not all of it arrives. This method is called **forward error correction** because the redundant data—the error correction information—is sent in advance of any known problems.

The redundant data needed for forward error correction takes up space

Forward error correction requires lots of redundant data because there is no way to tell when error correction will be needed. And that redundant data takes up space. On a compact disc, error correction is needed to compensate for possible scratches and minor manufacturing flaws. The older 63-minute compact disc format can actually store about a gigabyte (1024 megabytes) of information; but depending on the correction mode, up to 46 percent of the space is given over to error correction. The more robust the error correction, the more space it needs.

Compact discs employ one of two error correction modes: mode one has more robust error correction, using up almost 46 percent of the space and leaving 555 megabytes for usable data on a 63-minute CD (650 MB on the more common 74-minute CD). Mode two, with less elaborate error correction, uses approximately 38.5 percent for correction, leaving 630 MB for usable data (740 MB on a 74-minute CD). Publishers use mode one for CD-ROM discs containing software and other computer data, and mode two for discs where the occasional error is not fatal, such as those containing predominantly audio and video. Error correction for compact discs is done entirely within the player; the computer sees only a corrected data stream.

Forward error correction is also used for digital satellite video

Digital video broadcasts by satellite to modern small-disk satellite systems also employ a forward error correction scheme. The standard data rate for a satellite transponder is 40 Mbps. Of this stream, only a little more than half—23 Mbps—represents the active data stream; 17 Mbps (42.5 percent) is the redundant data stream for error correction. Among other things, the error correction in a satellite transmission protects against signal loss from a rainstorm, although very heavy rainfall will still disrupt the signal.

Digital bandwidth values should not be taken at face value

With forward error correction, what is the correct value for the bandwidth of a digital satellite video transmission? Arguably the bandwidth should be specified as 40 Mbps because that's what the transponder is sending; certainly the components in both transmitter and receiver must be capable of handling 40 Mbps. Or perhaps it should be 23 Mbps, because that's the useful data rate, and most times the error correction information isn't needed. This illustrates some of the problems in specifying bandwidth—and why bandwidth figures should not be taken at face value, particularly in the digital world.

Some digital signals require compression

Compression and quality Other problems arise from limitations in digital bandwidth. Because bandwidth costs money, the only practical way to send high-bandwidth information, such as video and sound, requires compressing the data stream, thus cutting down the number of bits per second sent through the channel. However, compression can compromise the signal quality mildly, or, depending on the type and amount of compression, seriously.

There are two basic types of compression, **lossless compression** and **lossy compression**. In lossless compression, lengthy patterns of repeated data are replaced with more compact ways of representing them precisely. This sort of compression is lossless because the expansion process can return the data stream to the exact original. For instance, in this book, a lossless compression algorithm might replace the word "bandwidth" with the number one. Since the word "bandwidth" is nine bytes and the number one is a single byte, a nine-to-one compression ratio would be achieved. The expansion process would go backwards, replacing all instances of the number one with the word "bandwidth." Lossless compression is essential for any situation where losing even a single bit can be fatal, as with a computer program. But with most material, lossless compression can't achieve the same compression ratios as lossy compression.

Lossy compression, on the other hand, works best when the loss of a few bits makes little or no difference to the end product. Lossy compression schemes are thus generally used for data like images and sounds, where the loss of a few bits results only in lower quality. As an example of how lossy compression works, imagine a digital picture of a field of red tulips. There are many different shades of red in a tulip, so a lossy compression scheme might reduce the number of shades, replacing some of the shades with others that are extremely similar. Because each different shade takes up space, reducing the number of them in the picture reduces the file size significantly (sometimes by hundreds of times). Of course, the quality of the picture worsens as well, but lossy compression algorithms generally try to take advantage of limitations in the human eye or human ear to remove information that we're unlikely to notice in a picture, sound, or video.

Expense of digital Digital has additional problems related to cost. First of all, some digital devices are relatively new and expensive, as anyone who has shopped for the latest digital gadget can attest. Second, some of the advanced forms of digital bandwidth described in this book are not yet available; others are available, but only at very high prices.

Some kinds of digital equipment are new and expensive, but prices will come down

However, one of the certainties is that the cost of computing devices and digital bandwidth will drop steadily. Already, some compact disc players sell for below $50 and offer far higher quality than analog record players that used to cost twenty times as much. Similarly, the cost of high-speed digital connections has dropped in recent years and will continue to drop as the cost of the electronic equipment used by the telephone companies falls and as the demand for the high-speed lines increases.

Political and social problems Some of the problems facing digital bandwidth are more political and social than they are technical, and these problems can lead to higher costs for digital information than would seem reasonable. The many parties who develop digital devices do not always agree on how to handle information, and this disagreement results in the development of often-expensive translation and conversion systems. Computers are much more difficult to set up and use than radios and televisions, and this leads to higher collateral costs for setup and maintenance.

Lack of agreement on standards, government regulation, and unresolved copyright issues can also cause problems

Organizational and legal issues also come into play. Older companies that are accustomed to traditional ways of publishing and distributing information may not handle digital information effectively or appropriately, reducing the advantages of working with digital information. Of course, the same can be said of new companies set up to publish information electronically; newness does not imply an understanding of digital information.

Government has also become involved in many activities that relate to digital bandwidth, through standards setting, broadcast licensing, spectrum sales, and regulation of pricing and content. Some of the main advantages of digital signals raise, at the very least, questions about copyright law. When anyone can make a perfect copy of a digital signal and transmit it to someone else without needing an expensive high-bandwidth pipe, the accepted rules may have changed even if the relevant laws have not.

Benefits of Digital

Of course, although there are some sticky problems with digital bandwidth, its advantages are significant. Because of those advantages, digital systems are supplanting analog ones in many places. Just as the digital CD virtually eliminated the analog LP, digital video will undoubtedly take the place of analog, and for many of the same reasons.

Using a narrower bandwidth to send signals takes longer, but may be beneficial if real time is not an issue

Trading bandwidth for time Digital bandwidth is in one sense elastic; you can easily trade time for bandwidth. Any digital signal, however large, can be sent in any bandwidth, as long as it is not synchronous (that is, it does not have to be sent in real time). Video conferencing must be synchronous; but a video clip for later viewing is not time-dependent. A one-hour digital video might take days or even weeks to push through a narrow bandwidth pipe, but it will get through intact. It doesn't matter if the digital bandwidth is narrow or wide; once a signal has been stored in digital form, it can be sent at a slower or faster pace to take full advantage of the bandwidth available. Furthermore, any interruption in the signal is easily managed; once the remaining bits are sent, the receiving device can stitch the pieces together perfectly and automatically.

Perfect copies In analog systems, copying always leads to signal degradation. In properly functioning digital systems, a signal can be copied or relayed perfectly. Being able to make perfect copies provides numerous benefits. For instance, a digital signal can be modified without fear of adding generational degradation, something that's not true of analog signals. Also, digital information can often be copied and transmitted much more cheaply than analog information of the same quality level, which would require expensive high-end equipment to minimize signal loss.

Digital signals can be copied without the addition of noise or distortion

The ability to make perfect copies of digital information is also excellent for making backups. If one copy of a digital original is lost, a backup can replace it exactly. In contrast, if an analog master to a sound or video recording is lost, there's no way to recover the original level of quality. However, this capability also raises numerous issues surrounding copyright law—the fact that you can make a perfect digital copy of an audio CD doesn't mean that you're legally allowed to do so. In the past, although copying has been a concern, the loss of quality inherent in recording a long-playing record or photocopying a printed page has always reduced the concern held by the creators of the content. Now, however, anyone with inexpensive equipment could set themselves up making and selling exact copies of other people's work.

Digital copies are perfect backups, but this perfection can also lead to copyright abuse

Intermixing Whereas analog signals generally require a specific circuit or portion of the spectrum for each type of signal, digital signals can be intermixed at will. Because digital signals are all made up of bits, anything that can be converted to digital format can be transmitted and stored digitally, and a computer or other electronic device will do the work of recreating the digital bits as an exact copy of the original. For instance, using a digital connection to the Internet, you can download audio and video files at exactly the same time that you're receiving text-based e-mail. This fact enables digital connections to transmit many types of

Digital signals can be intermixed

data and digital computers to manipulate those different types of data, something that isn't possible in the analog world, where transmission and playback devices are generally single-purpose.

Not all digital devices feature every one of these advantages. Audio CD players, for example, cannot understand non-audio data on CD-ROMs. DVD disks and players are deliberately designed to foil counterfeiters. And VCRs that can copy digital TV signals without degradation remain wildly expensive commercial devices. But almost any computer hooked up to the Web can take advantage of these digital benefits, and many more devices will in the future. The simple fact is that because digital information can all be reduced to zeroes and ones, it's far more flexible to work with most data in digital format.

In Conclusion

The terms surrounding bandwidth are bandied about in many ways. As we have seen, analog bandwidth can be particularly confusing, because there are often several different ways of defining a particular type of bandwidth. Digital bandwidth can be easier to master once we understand the fact that there are eight bits in a byte and that digital bandwidth is expressed in bits per second, whereas digital storage is generally spoken of in terms of bytes. You should have a grasp of the problems and benefits inherent in analog and digital information.

At the start of this chapter, we looked at how audio bandwidth is measured. In the next chapter, we'll see how bandwidth affects the various ways of broadcasting audio in formats ranging from AM radio all the way to digital satellite audio.

Chapter Five

Broadcast Bandwidth: Audio

Digital advocates believe that widespread replacement of analog broadcasting with digital broadcasting is inevitable. For one thing, they say digital broadcasting can use bandwidth in the radio spectrum more efficiently. But analog receivers—both radios and televisions—are so inexpensive and common that digital broadcasting may never completely replace analog broadcasting. In the third world, where costs are much more critical than in more developed countries, analog broadcasting may well survive forever.

Analog and digital broadcasts are likely to coexist

The information sent over broadcast bandwidth comes in several different forms. All of them apply to radio broadcasting, so radio is the main topic of this chapter, with a little about the audio component of television.

Here is a list of the standard broadcast types, both analog and digital:

- Audio information in analog form—standard radio and television audio
- Stereo audio information that enhances the main analog monaural audio—standard stereo radio and stereo television audio
- Secondary audio information that is separate from the main audio program—the Secondary Communication Authority (SCA) service that enables the transmission of separate programming via FM radio and the secondary audio program in TV audio
- Digital audio that carries the same program content as the main audio—the proposed AM and FM digital audio formats
- Digital audio that carries program content different from the main audio—a proposed adjunct format for television
- Digital data that is directly related to the main program—station identification, program information, closed captions, and so on
- Datacasting or digital data broadcasting that is unrelated to the main program—SCA digital services in FM and several digital services on television

The following sections examine each of these major audio broadcast formats to see how they use their allotted spectrum range to deliver bandwidth.

AM Radio

AM broadcasts remain low in quality because they do not use all of their allotted bandwidth

AM radio operates in a portion of the broadcast spectrum, roughly 525 kHz to 1705 kHz, that is sometimes called "medium frequency." (See the Frequency Ranges table on page 70.) AM radio stations were the first broadcast stations, long before high-fidelity sound. Seventy years ago, AM radio

Modulation and Carrier Waves

A **carrier wave** is a radio wave whose amplitude or frequency is varied, or **modulated**, to encode information. Modulating the signal strength, or the height of the radio wave, is called **amplitude modulation**, better known as AM.

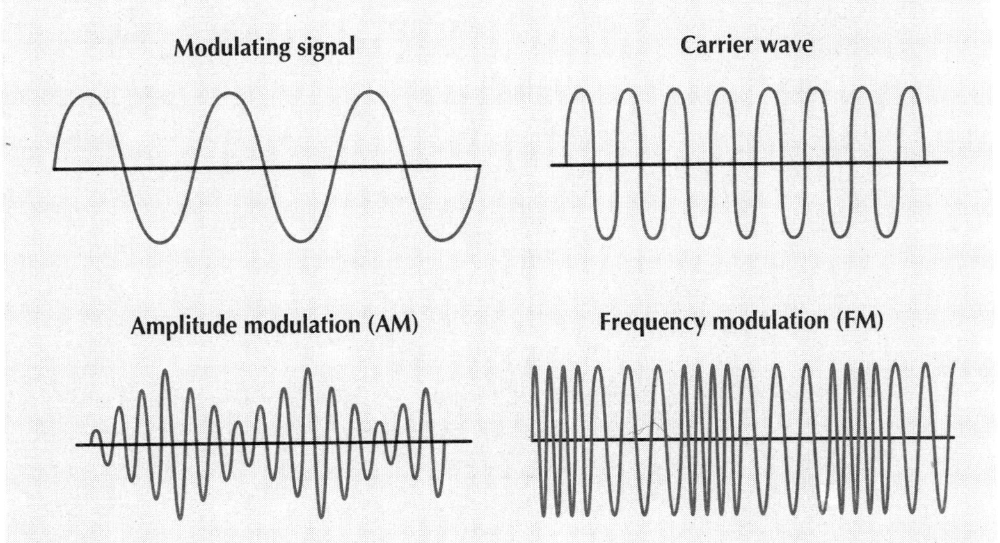

Modulating the frequency of the wave is called **frequency modulation**, or FM. If you imagine a sine wave, amplitude modulation would vary the height of the peaks and the depths of the troughs, whereas frequency modulation would vary the number of peaks and troughs in a given amount of time.

had a bandwidth of roughly 3 kHz with a pronounced loss of high-frequency sound. AM spectrum assignments actually allow for 10 kHz of bandwidth; by the 1950s, the better stations and radios were using 4 or 5 kHz. Despite the fact that using more bandwidth would allow for improved sound quality, no one used the full amount. AM broadcasters don't bother to improve their sound quality, arguing that people who want better sound listen to FM radio. Even the AM tuners in expensive stereo equipment are almost always poor

Frequency Ranges

Frequency	*Description*	*Wavelength Range*
3–30 kHz	Very low frequency	100,000–10,000 m
30–300 kHz	Low frequency	10,000–1000 m
300–3000 kHz	Medium frequency	1000–100 m
3–30 MHz	High frequency (short-wave)	100–10 m
30–300 MHz	Very high frequency	10–1 m
300–3000 MHz	Ultrahigh frequency	1 m–10 cm
3–30 GHz	Superhigh frequency	10–1 cm
30–300 GHz	Extremely high frequency	1 cm–1 mm

in quality; very few have **flat response** to even 5 kHz on AM, even though they are flat to 20 kHz in all other circuits. So, as a result of the expectations of broadcasters rather than any technical limitation, AM radio remains poor in quality.

High-quality AM radio in New Zealand is the exception

There is always an exception. In New Zealand, high-quality AM radio developed because FM broadcasting did not begin until 1982. The New Zealand–made McKay-Dymek AM tuner has become legendary among serious audiophiles as the best AM tuner ever built. New Zealand does have the major advantage of being isolated, with relatively few AM stations; **interference** between stations—which discourages wide-bandwidth broadcasts—is a bigger problem everywhere else.

Countries use different parts of the spectrum for AM

AM broadcasting takes place in several other areas of the radio spectrum. In Europe and North Africa, the long-wave band, from 155 to 281 kHz, remains in use for AM broadcasts. Around the world, many stations broadcast AM radio in 13 very crowded short-wave bands from 2.3 to 26 MHz. And the television system in France uses AM for audio.

There are about 5000 AM stations in the United States. AM radio is susceptible to interference both from other stations and as a result of natural and human activities. AM broadcasts can travel over 1000 miles at night, which can cause interference unless the station has been assigned a clear channel so that no other station within a thousand or more miles uses the same frequency. Before the satellite era, a clear-channel station had the widest reliable reach of any single transmitter; short-wave radio transmissions do travel farther than AM radio, but with far less reliability.

Clear channel assignments can protect frequency assignments for more than 1000 miles

Medium-wave AM radio is an international format, although from region to region there are some minor variations in where the band begins and ends. Most of the world's AM stations transmit on frequencies that are set at 9-kHz intervals (531, 540, 549 kHz, and so on). In North and South America, the frequencies are set at 10-kHz intervals (530, 540, 550 kHz, and so on). This discrepancy can cause problems for digitally tuned radios, unless they are intended for international use and therefore designed to switch between the two frequency intervals.

Standard intervals between frequencies vary internationally

Even though most sound systems are stereo, AM radio is still generally monaural. The abortive introduction of AM stereo in the United States in 1982 shows what happens when broadcast standards are left entirely to the marketplace. The FCC did not specify a single AM stereo standard; instead, four incompatible systems were allowed to compete in the open market. The resulting confusion killed AM stereo. Even though circuits that could switch automatically between systems eventually became available, only a handful of stations and radios bother with AM stereo today. Internationally, the few countries that have AM stereo have set their own standards.

Without a single AM stereo standard, incompatible systems caused confusion

An AM signal takes up twice the spectrum space—twice the bandwidth—necessary for the information it carries. Even though a 710-kHz signal occupies a span from 705 kHz to 715 kHz, it is really only 5 kHz wide, as you can see in Figure 5-1. The broadcast signal from 705 to 710 kHz—one sideband—contains the same information as the signal from 710 kHz to 715 kHz, the other sideband. A single sideband

Single sideband transmissions, common in point-to-point radiotelephones, are seldom used for radio broadcasts

(SSB) transmission suppresses one sideband and allows twice as many stations in the same spectrum space. SSB is common in both point-to-point radiotelephones that use AM modulation and in amateur radio. However, SSB requires a more complex and expensive receiver and is rarely used for radio broadcasts, with the exception of a few short-wave programs. One use of the second sideband was in the abortive AM stereo transmissions, which achieved stereophonic sound by modulating the two sidebands differently to send stereo information.

Figure 5-1 ***AM radio duplicates the signal in each sideband of its 10-kHz allocation.***

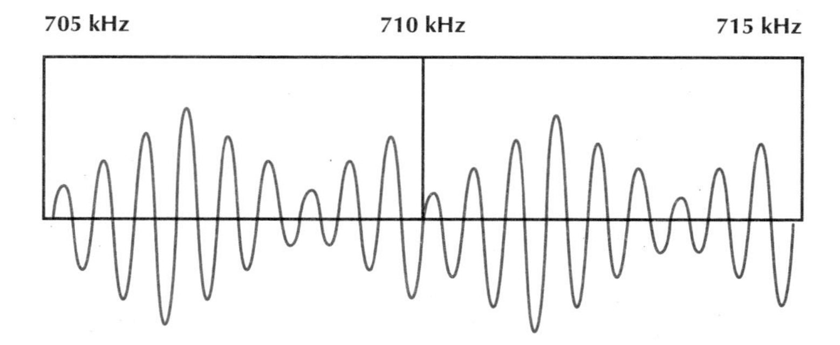

Despite the ability to send digital audio over the AM radio band, its use seems unlikely

Despite limited bandwidth, it is possible to send digital data over the AM radio band. USA Digital Radio, a private corporation, has proposed a format called In-Band, On Channel (IBOC) that features a raw data rate of 128 Kbps, with 96 Kbps of usable bandwidth after **forward error correction**. This is enough for decent-quality compressed monaural sound, with 2.4 Kbps of this bandwidth allocated as a small data stream for station and program identification and other related non-audio information. AM-IBOC was intended to duplicate AM analog audio in digital form, not to send an independent data stream. However, USA Digital Radio pulled AM-IBOC from field tests in May 1996 because of sound quality, so FCC approval seems highly unlikely. International standards for digital audio over AM seem unlikely as well.

FM Radio

In the first three decades of FM broadcasting in the United States, revenues and profits were meager. In 1955, the FCC authorized SCA subcarriers for transmitting background music and the like to help give FM stations additional income. In the late 1960s, FM radio began to gain listeners rapidly, pulling ahead of AM stations. By the late 1970s, FM stations were ahead in both market share and advertising revenue. Today, there are about 6000 FM stations in the United States. Many cable television systems carry FM radio stations, sometimes including stations outside the local area.

Approximately 6000 FM stations exist in the United States

FM radio works quite differently from AM. The audio signal modulates the frequency of the FM carrier, rather than the carrier amplitude as in AM radio. An FM radio locks on to, or captures, the FM carrier and rejects any other signal that does not follow the carrier, including most kinds of noise and any weaker FM stations on the same frequency. This key advantage is why you almost never hear two FM stations at the same time, a common problem with AM radio. The entire FM band spans 20 MHz, from 88 to 108 MHz, compared to only a little more than one megahertz for the AM band. Each FM station gets 200 kHz of spectrum, a wide bandwidth that supports a signal extending to nearly 100 kHz. This extra bandwidth is used to provide a better **signal-to-noise ratio** and several additional subcarrier signals. In effect, an FM station behaves like several radio stations in one.

FM signals modulate frequencies; each station is allotted 200 kHz of spectrum

Here's how the FM signal breaks down: the main monaural audio signal is transmitted from 0 to 15 kHz. If the station is transmitting in stereo, there is a subcarrier signal (another modulated signal transmitted with the **baseband** audio) at 19 kHz. This subcarrier turns on the stereo circuits in an FM tuner and lights up the stereo indicator. From 23 kHz to 53 kHz, the station transmits the stereo difference signal—the difference between the left and right channels. The receiver

Stereo transmissions include a stereo difference signal

The Parts of an FM Channel

Portion of the Channel	*Usage*
0–15 kHz	Main monaural audio signal (R+L)
19 kHz	Stereo indicator subcarrier
23–53 kHz	Stereo difference signal (R–L), centered at 38 kHz
57 kHz	RBDS subcarrier
67 kHz	SCA subcarrier signal
92 kHz	Second SCA subcarrier signal

extracts the left and right channels by adding and subtracting the monaural signal and the stereo difference signal.

The stereo difference signal is used to separate the left and right stereo channels

Although "adding and subtracting" sounds funny, that's precisely what happens. The monaural signal consists of the right and left channels mixed together, or R+L. The stereo difference signal, which is the difference between the signals, is thus R–L. If you just added the stereo difference signal to the monaural signal, you'd be left with only the right channel (R). On the other hand, if you just took the difference between the two channels, you'd have only the left channel (L). But if you do both, you end up with both the right and the left channels, nicely separated.

Most SCA audio subcarrier signals are available only by subscription

Not all FM stations broadcast subcarriers, but many do. A second subcarrier at 67 kHz is the third signal. It carries the SCA signal, which can contain a lower-quality audio program in mono, such as background music, readings for the blind, or other programming. A third subcarrier at 92 kHz is the fourth signal, another SCA signal. In most areas, the SCA signals are private, available only to subscribers who get a special SCA radio as part of a subscription to a service. In fact, under U.S. law, it may not be legal to monitor certain SCA transmissions. SCA-capable FM radios are not normally sold on the open market.

An SCA subcarrier can also carry digital data. Three companies offer mutually incompatible protocols and hardware for transmitting data over FM radio. The data rates run from 8 Kbps to 19.2 Kbps; stations that want to transmit SCA audio simultaneously are limited to the slower speeds. To receive the data, you need a computer accessory with an FM tuner that captures and stores the data on the computer's hard disk drive. The data that is sent in this fashion is usually unrelated to the FM programming. Sending the same data, such as stock quotes or constantly updated sports scores, to a wide audience is often called **datacasting**.

An SCA subcarrier signal can transmit digital data

Yet another subcarrier at 57 kHz conveys the Radio Broadcast Data System (**RBDS**), a digital signal with a relatively slow bandwidth of 1187.5 bps. RBDS sends information that has both public and private portions. A radio capable of receiving RBDS typically uses a small display to show the publicly available information—the station call letters, the name of the program or music that is currently playing, traffic reports, and emergency warnings. Several nonpublic portions of the RBDS signal can be used for any purpose. There's enough capacity to handle several thousand pagers or to send data to remote-controlled billboards—perhaps a scrolling display mounted on the side of a bus. The SCA audio and datacasting methods leave space for RBDS.

RBDS subcarrier signals are digital, with both public and private portions

Upcoming Digital Radio Formats

A variety of proposed systems would send digital audio with the analog audio in the FM band, with the goal of providing "CD-quality audio" along with other potential data services. One set of proposals, which has since been pulled from field testing due to interference problems with adjacent stations and analog receivers, is FM In-Band, On Channel (FM-IBOC). Although the digital signal could also have been used to send a separate program, FM-IBOC carried digital sound that duplicated the FM analog program. Depending on the specific proposal, FM-IBOC transmitted between

FM-IBOC was pulled from field testing due to interference problems

128 Kbps and 256 Kbps to provide a stereo signal with compression. FM-IBOC was in the FM band but didn't use the same channel; it left the present 200-kHz FM channel intact and used bandwidth in the adjacent channels above and below the assigned frequency. Thus a station assigned to 98.1 MHz would have transmitted FM-IBAC at 97.9 and 98.3 MHz. Two stations never occupy these adjacent frequency assignments in a given area, although stations are sometimes assigned alternate channels two steps away. (In the above example, another station might be at 97.7 or 98.5 MHz.)

FM-IBAC effectively reduced interference, but its bandwidth is limited

A slightly different proposal is FM In-Band, Adjacent Channel (IBAC). Whereas FM-IBOC used the first adjacent channel (97.9 and 98.3 MHz are the first adjacent channels for 98.1 MHz), FM-IBAC instead uses a second adjacent channel (either 97.7 or 98.5 MHz) to provide 160 Kbps of bandwidth. In field tests, the latter method has proven more effective at reducing interference both in the digital signal and with nearby analog FM signals. 160 Kbps isn't really enough bandwidth for high quality stereo sound, so it remains to be seen if FM-IBAC will succeed.

"FM Digital" would replace the SCA and RBDS subcarrier signals, but its bandwidth is also limited

Another, somewhat unusual method, rather ambiguously called "FM Digital," replaces the SCA signals and RBDS signal with a single 240-Kbps digital signal and an anticipated **throughput** of 160 Kbps after forward error correction. The equivalent of the RBDS signal would be placed in the data stream. 200 Kbps is a little shy of quality stereo audio, but if the broadcaster were willing to dump the analog stereo signal and transmit only analog mono, FM Digital could carry 400 Kbps, enough for good quality stereo sound. Unlike FM-IBOC, the FM Digital signal stays within the channel limits allowed by the FCC and may not require FCC approval. FM Digital has not yet run through full field-testing.

The Limitations of FM Radio

FM radio quality suffers from several problems, some inherent to the medium, and others a result of broadcast practices. Noise is the most common problem. One reason is a weak signal, either because you are too far from the transmitter or because you are inside a steel frame building that acts as a shield against the signal.

Common limitations include weak signals and multipath problems

But the most common cause of noise is the signal's reflections off buildings or mountains. In this case, both the original signal and the reflected signals ("multipath signals") converge on the radio. A stationary radio can usually cope because an FM radio latches onto the strongest signal. In a car, though, multipath creates severe noise problems as the radio switches from direct and reflected signals, sometimes several times a second. An FM diversity receiver tries to overcome these problems with two antennas; the radio constantly monitors both antennas and selects the antenna with a better signal.

With a clean FM signal on a stationary receiver, sound quality still falls far short of what FM can deliver. There's a lifeless quality to the music broadcast by many FM stations. A good FM broadcast could, and should, sound nearly as good as an audio CD; the slightly higher distortion and smaller **dynamic range** don't matter for most program material. Most FM stations either improve or damage their sound quality—depending on your point of view—by processing their audio to make the station sound louder; the processor includes a limiter to prevent overmodulation. The same kind of processing carried further makes commercials sound louder.

Audio processing affects sound quality

A few stations do take the care to broadcast clean, unprocessed sound. Ironically, competition seems to foster poor sound, because stations often use audio processing to sound louder and thus easier to find when someone is scanning for a station. In Europe, government-operated FM stations that are relatively free of commercial pressure are

Audio is processed to make it sound louder

more likely to exploit FM's full sound capabilities. The definition of good sound, however, is open to question. For someone listening in a noisy car, the louder compressed and limited sound is arguably better than the wider **dynamic range** of unprocessed sound.

A louder signal with less data is sometimes preferable

Just as the definition of good sound is open to question, so too is the question of whether an FM station uses more or less bandwidth with this technique of increasing the volume. Although less information reaches the listener when the signal has been processed and limited, the listener might hear even less of the signal—for example, a pianissimo passage in a car—if that processing hadn't taken place. In short, you sometimes must take local conditions into account when evaluating bandwidth, especially analog forms destined for the human ear or eye.

International Differences in FM

FM broadcasts are not standardized around the world. Although the 88–108 MHz band is the most common, Japan uses 76–90 MHz and Eastern Europe mostly uses 66–73 MHz. In North America, FM stations are on the odd tenths (88.3, 88.5, etc. MHz) while elsewhere in the world, stations are on even tenths (88.0 MHz, 88.2 MHz, and so on), odd tenths, and sometimes finer fractions. FM tuners sold for international use can tune every tenth of MHz or finer steps; radios sold in Japan usually tune from 76 to 108 MHz. A very few radios tune from 66 to 108 MHz. FM radios with extended tuning can pick up the audio portion of television broadcasts in most countries. Finally, North American FM broadcasts have slightly different high-frequency equalization from the rest of the world; your tuner is adjustable if it has a switch labeled 75 microseconds (North America) and 50 microseconds (rest of world). Stereo FM works the same way around the world.

Satellite Audio

Currently, audio broadcasting by satellite is done as an adjunct to satellite television, a topic you will learn about in the next chapter. Because these audio broadcasts require the receiver to have a satellite dish at least 18 inches in diameter, these services are for fixed locations only. Ku band satellites broadcast digital video and also carry audio programming, which is typically a stereo signal sent as a 384-Kbps compressed digital data stream that is converted to analog audio by the satellite receiver. The programming, usually in a specific music genre (country, jazz, and so on), may be from a terrestrial station or a service set up for the satellite.

Ku band satellites can deliver audio programming to fixed locations only

Digital Audio Radio Service

A more ambitious form of digital satellite radio, called Digital Audio Radio Service (DARS), seeks to supplement or replace the present analog AM and FM radio with direct satellite broadcasts. Again, 384 Kbps is the likely bandwidth for a stereo program, but because the signal is digital, its bandwidth can be adjusted easily, so talk radio programs could use less bandwidth and special concerts could use more. The capability of varying the bandwidth used based on the content is one of the main advantages of digital bandwidth.

DARS could replace analog radio

In the United States, the FCC has authorized digital satellite radio on the S band (2310–2360 MHz), which would be easily received without the need for a fixed satellite dish. Will the S band radios be small enough to fit in your pocket? No one knows for sure. Even with S band, car radios are a big part of the radio audience, so tunnels, underpasses, and tall buildings pose a serious problem to mobile listeners because satellite signals do not penetrate such barriers. Because the satellite will be positioned over the equator, in North America, obstacles to the south of the receiver will block the signal.

Digital satellite radio, authorized on the S band, may need supplemental ground transmitters to be truly effective

To serve mobile listeners, broadcasters propose supplementary transmitters on the ground that could relay the satellite programming to areas the satellite signal cannot reach. Such transmitters raise interesting questions, however. S band radio is intended to be a very wide-area service, perhaps even national. But a local supplementary transmitter might well carry local traffic announcements and perhaps local commercials. Would these transmitters develop into full-fledged terrestrial stations over time? Or would purely national radio programming truly be in the best interests of the country?

DARS supporters argue more service for more people

Proponents of DARS argue that it would provide CD-quality sound, increased programming diversity with programming aimed at niche as well as mass audiences, additional services such as car navigation assistance, and radio for the part of the U.S. population not served by radio.[1] In addition, some jingoistically argue that DARS in the S band serves the national interest because other countries are working in the L band.

DARS opponents fear homogenized content and the destruction of existing stations

On the other side of the argument, opponents of DARS (primarily associations of broadcasters and over 100 individual radio stations) say that the system would result in less diverse programming due to the national audience. They also fear that DARS would destroy existing terrestrial radio stations, many of which are currently barely afloat financially. That, they say, is not in the public interest, because existing radio stations carry largely local programming. Finally, there are questions regarding the quality of the sound that will be unanswered until more tests are done.

Two DARS licenses were granted for a total of $173,234,888

In April 1997, the FCC auctioned two licenses for the spectrum from 2320–2345 MHz. Because the FCC had allotted only a total of 25 MHz for DARS, there was room for only two licensees, given that 12.5 MHz is necessary for

1. According to one study, 0.3 percent of the population has no radio service, but that study failed to account for AM radio, which can travel great distances.

a viable DARS system. Four applicants vied for the licenses, and the price tag was high: Satellite CD Radio, Inc., won the first license with a bid of $83,346,000 and American Mobile Radio Corporation won the second with a bid of $89,888,888, for a total of $173,234,888. Obviously, bandwidth is worth a lot of money to the right people.

However it works out, DARS isn't likely to affect many people in the near future. The FCC expects the first service to begin sometime in 1998 or 1999, and based on adoption rates of other technologies (such as CD players in cars), it expects that the overall penetration of DARS may be no more than 4 percent by the year 2005. Although improved sound quality is the primary reason given for the need for DARS, it seems likely that the significant bandwidth and coverage it offers may make it more useful for other applications such as transmission of traffic, weather, and emergency advisories directly to vehicles, especially in remote areas that wouldn't otherwise be served by that sort of radio broadcast.

DARS may prove to be most useful in serving moving vehicles and remote areas

Canadian Satellite Radio

Meanwhile, north of the border, the Canadian government is planning to authorize digital satellite radio on the L band (1452–1492 MHz), following the Eureka 147 digital stereo format developed in Europe. But in the United States, the L band is used in aeronautical telemetry communications for flight testing and other military applications, and the Canadian military uses the S band. (In fact, the S band is also used for aeronautical telemetry communications in the United States, but the organizations that rely on it say that moving operations out of the S band would be much easier than moving them out of the L band.) It's not clear how the conflict between the two competing proposals will be resolved. Needless to say, satellite-broadcasting coverage cannot be neatly stopped by an arbitrary international border.

Satellites do not respect geographic borders

The proposal for L band has a lot of support

The Canadian proposal has a number of advantages. The lower-frequency L band may need fewer supplementary transmitters, and the Canadians claim that L band digital radio has been proven to be interference-free, even in moving cars and personal portable radios. Also, most other countries seem to be jumping on the Canadian L band-wagon, so an international standard could speed development, increase coverage, and reduce hardware costs in the long run. The Consumer Electronics Manufacturers Association (**CEMA**) has also come out against the use of the S band, mostly because it's not clear that digital radio in the S band can be received acceptably in a car. The FCC feels that CEMA's technical complaints were not based on appropriate testing and discarded them when deciding on the S band over the L band.

Unlike the United States, Canada plans to cease analog broadcasts

One major difference is that the Canadian proposal also foresees the cessation of AM and FM broadcasts within Canada by the year 2010, with all existing stations moving to the L band. Canada is explicitly discouraging the use of AM and FM in-band digital audio. In contrast, the U.S. broadcast industry expects to continue analog broadcasts indefinitely, and the FCC has said that it remains open to proposals like FM-IBOC, FM-IBAC, and FM Digital that would transmit digital radio in the existing FM band. If the current plans come to fruition, the United States and Canada will for the first time diverge significantly in broadcast practice. Future radios may not work in both countries, aside from the spillover signals across the border.

The United States may once again go its own way

This situation would seem to be primarily a case of differing national interests; it's instructive to see how governments view their role in bandwidth allocation. In many of these cases, such as with cellular phones, the United States ends up backing a standard that's supported nowhere else in the world and getting away with it because of our huge market for consumer electronics. Although the FCC hasn't made any decisions yet, digital radio may be yet another case in which the United States and the rest of the world diverge.

In Conclusion

In this chapter, we looked at the different forms of radio bandwidth used for audio, focusing on the two most common, AM and FM. AM radio is notable for its minimal bandwidth, whereas FM radio has enough to include not only audio, but also subcarriers that handle extra information, including digital data.

A number of other proposals for transmitting digital audio via FM have appeared, although none has proved tremendously promising in field-testing. The real battle is slated for higher in the spectrum, with Canada and much of the rest of the world settling on the L band at 1452–1492 MHz and the United States auctioning licenses for digital satellite radio in the S band at 2320–2345 MHz.

In the next chapter, we'll turn our attention from audio to video.

Chapter Six

Broadcast Bandwidth: Video

Television has become the major broadcast medium for entertainment and sometimes information in the past fifty years. In the beginning, televisions were novel and usually installed as the centerpiece in living rooms. When televisions proliferated in the 1960s, many households acquired two or three sets, which had become just another piece of furniture. In the past decade, some televisions have once again become the centerpiece, at least for the wealthy and the "television-involved," only now as a large-screen projection television with surround sound in a home theater.

What's amazing about this 50-year progression is that the basic television broadcast standard has not changed since the very beginning, except for the addition of color in 1953 and stereo sound in 1984. The quality of television cameras and television sets has improved greatly, but the picture structure and bandwidth remain the same. A signal identical to one sent in the 1940s could be captured and played on any television set today, and a working 1940s set could receive today's shows in black and white.

Television standards have not changed much

This chapter looks in detail at today's television broadcast and display systems. After the basics are covered, the focus turns to satellite television, which can be either analog or digital, and **datacasting**—ways of transmitting digital information within the television signal. The chapter concludes with a brief look at high-definition television, which requires much higher bandwidth but promises higher-quality pictures and sound, and motion pictures.

Television Bandwidth

As you saw in the "Video Bandwidth Measurements" section in Chapter 4, video bandwidth can be measured in a number of different ways, including the number of lines of resolution, the bandwidth of the broadcast **channel**, and the bandwidth of the video signal within that channel. Just as portions of the FM radio signal are used for non-audio information, portions of the television broadcast channel are used for non-video information.

The TV Picture

Overscan hides flaws

In the North American television system known as **NTSC** (National Television Standards Committee), a video frame has 483 visible lines—if you can see the entire transmitted picture. But nearly all television sets mask out part of the picture (called **overscan**) to hide image flaws. The earth's magnetic field can rotate the image slightly, for example, and the picture size and position change as the set ages.

Interlaced scanning uses two alternating fields to produce 60 frames per second

In essence, television works by displaying many still images one after another, just as children use flipbooks to animate simple drawings. To convey motion, 30 frames are sent every second. A 30-frame-per-second image would flicker badly, but sending 60 complete frames per second would require too much bandwidth, so each frame is broken into two "fields," one consisting of the odd-numbered lines and the other the even-numbered lines, a technique called

interlaced scanning. Each field amounts to 241.5 lines and ends in a half line. Because each field is equivalent to half a frame, and 60 fields or 30 frames are broadcast in a second, motion is conveyed in half the bandwidth that 60 full frames a second would take. Unfortunately, the resulting interlaced picture is not as good as it might be.

Interlaced scanning: odd + even = full picture. ***Figure 6-1***

Odd + **Even** = **Full picture**

Progressive Scanning

There is an alternative to interlaced scanning. Sending every line sequentially is called **progressive scanning**. Progressive scanning is not used in any broadcast television today but is proposed for future television standards. Nearly all computer displays use progressive scanning. The problem with progressive scanning is that it requires more bandwidth than is available in broadcast television; when there's plenty of bandwidth available, as with a computer display, progressive scanning can provide a much higher quality image than the interlaced scanning in even the best television set.

This fact accounts, in part, for why computer monitors are so much more expensive than similar-sized television sets. Numerous products have appeared over the years that enable you to use a television set as a display for a computer, but no matter what technical tricks they employ, the resulting display is always much fuzzier than even an inexpensive computer monitor, especially for text. Because most information on the Internet is text, fuzzy display is one of the major problems faced by the products that bring Internet access to your television via a set-top box.

TV Broadcast Bandwidth

Television transmissions use amplitude modulation and are allotted 6 MHz of bandwidth

The bandwidth for a television channel is 6 MHz. The video portion of the signal is sent in **amplitude modulation**. Unlike AM radio, which duplicates the 5-kHz signal in the 10-kHz AM band, the video is sent single sideband to save spectrum space (and thus bandwidth), although 1.25 MHz of the unneeded sideband is left in the signal. That 1.25 MHz has been wasted bandwidth for decades, but was left in during the planning for television in the early 1940s to make television tuners easier to build. Today, there are some efforts to use this vestigial sideband for transmitting data, which will be discussed in the "Data in Video Sideband" section later in this chapter.

Monaural television audio can be heard on some FM radios

The audio portion of a television transmission is sent as an FM signal, and it can be received by FM radios that have an extended tuning range to reach the frequencies used by television channels. Even with extended tuning, FM radios

Broadband and Baseband

It's worth making a distinction between **broadband** and **baseband** when thinking about video bandwidth. A broadcast or cable television signal carries broadband video signals that contain many stations on the same medium (over the air or a coaxial cable) with up to hundreds of megahertz of overall bandwidth. Each station has its own 6 MHz of bandwidth, **modulated** on its own carrier frequency or channel assignment. The tuner in the television set or a cable box selects the desired program by tuning to a specific carrier frequency (channel).

In contrast, a baseband video signal carries only a single video program, including the sound. A television set accepts broadband signals at its antenna input, a baseband signal at its monitor or AV input.

The same distinctions can apply to a stereo system: a broadband signal would be the signal from the antenna to the tuner; a baseband signal would be the left and right audio signal, such as you would route between stereo components.

can receive only monaural TV audio; stereo FM radio and stereo TV audio are incompatible.

Many television stations also broadcast a secondary audio program (**SAP**) that is completely independent of the main audio. SAP can be used for a soundtrack in a second language or for a completely different program, in the manner of SCA audio for FM radio. Stereo VCRs will record the stereo audio along with the picture; many can be set to record the SAP on the linear track and the stereo audio on the helical hi-fi tracks.

SAP is independent of the main audio

VHF and UHF

Broadcast television takes place at several different locations in the radio spectrum. As with radio stations, TV stations in a given area do not occupy adjacent channels because of potential **interference**. However, channel pairs from adjacent frequency ranges—4/5, 6/7, and 13/14—can operate in the same area because of a frequency gap between each range. (The Frequencies of VHF and UHF Channels table shows the frequency assignments in 6-MHz increments.) In North America, channels 2–13 are in the **VHF** or very high frequency band (remember, these names were given in the 1920s, when 30 MHz was considered "high" frequency). Channel 1 was removed after World War II for government use.

Adjacent television channels would create interference

Frequencies of VHF and UHF Channels

Channels	*Frequency*
2–4	54–72 MHz
5–6	76–88 MHz
7–13	174–216 MHz
14–20	470–512 MHz
21–69	512–806 MHz

In 1963, the FCC required televisions to support UHF

When the supply of TV channels fell far short of the demand, the FCC opened up the **UHF** (ultra high frequency) channels 14–83 in 1952. But UHF grew very slowly because most televisions came only with VHF tuners; some TV set makers owned VHF stations and were reluctant to encourage competition by adding UHF tuners to their sets. Finally the FCC required all TVs to include UHF tuning in 1963.

The lower frequency channels cover more area

Eighty-two channels turned out to be an oversupply, so UHF channels 70–83 were removed from television broadcasting. The higher the frequency, the smaller the physical area a transmitter can reach; to compensate, UHF stations transmit with higher power than VHF stations. But VHF stations still cover more area, particularly the lower frequencies, such as channels 2 and 3.

Cable Television

CATV systems have blossomed into cable systems

Over-the-air television broadcasts depend on a clear, unobstructed path between the transmitter and receiving television antenna, with no mountains or large buildings nearby to produce reflected signals ("ghosts"). In many areas, clear reception is difficult or impossible, so neighborhoods grouped together to build a community antenna system (**CATV**), putting up an elaborate antenna to receive the signals with a cable signal feed to subscribers who shared the costs. Over time, many CATV systems blossomed into broader-scale cable systems, distributing not only nearby television stations, but other programming from around the country. Cable television runs past 70 percent of the U.S. population; about half of all television viewing is done with a cable connection.

Cables take over where signals are blocked by hills or other structures.

Figure 6-2

A cable system can carry many more channels

A cable television system works just like over-the-air television broadcasts. However, cable suffers less from interference problems than over-the-air systems; this fact means that many more channels can be made available. What's more, cable systems can use every channel, rather than being forced to avoid most adjacent channels, as over-the-air television must. Despite this, early cable systems only supported 12 VHF channels. Today, most cable systems offer 25 channels; more recent systems support from 50 to 150 channels. Newer 300-channel systems usually switch between two separate cables. Many cable services also carry FM radio.

Microwave Television

MMDS uses microwave frequencies and carries 33 NTSC channels

Television signals are also transmitted over the microwave frequencies 2.5–2.7 GHz. This band supports 33 channels of NTSC video. These television channels, called multichannel multipoint distribution systems (**MMDS**), and occasionally wireless cable, have been used in the past for limited-access video such as for a school system. Many MMDS transmitters are still reserved for educational broadcasting during school hours; they carry commercial services after hours. MMDS has been popular for pay television services such as movie channels, with about a million subscribers in the United States in 1996, although MMDS has lost some market share to the small-dish satellite receivers. The technical characteristics of MMDS are similar to VHF and UHF television, except that the MMDS transmitters are less expensive, typically low powered, and usually serve a relatively small area.

Color Television

Early color attempts were of poor quality and incompatible with monochrome sets

The original television standards in all parts of the world described only monochrome television. The first efforts to broadcast color sent three separate images (red, green, and blue) of equal quality in rapid succession. To fit in a 6-MHz channel, the three images were sent with lower quality (405 lines at 24 frames per second) than the monochrome image. This format, developed by CBS and first broadcast in 1951, was not compatible with monochrome television, and fared poorly because no one with a monochrome set could receive the pictures and very few people had color sets. CBS abandoned the system after four months.

Compatible Color Television

RCA's system, which separated the color and monochrome signals, became the NTSC standard

Meanwhile, RCA developed a compatible color system that did not make monochrome televisions obsolete. RCA did this by separating the color (**chrominance**) signals from the monochrome **luminance** signal, which remained at nearly full quality. The two chrominance signals with color information are transmitted with much lower bandwidth and higher noise. The lower-resolution color information takes advantage of the fact that the human eye responds more to brightness (luminance) than color when looking at details. However, the limited color information does restrict the overall picture quality. The RCA system for transmitting color was adopted as the NTSC system in 1953.

Composite Color Signals

The NTSC broadcast signal is a composite signal: the chrominance signals are superimposed on the luminance signal. The superimposition is possible because the luminance signal does not use the video bandwidth evenly. At multiples of the horizontal scan frequency, there is very little luminance information; the color subcarrier is placed in one of these multiples (3.58 MHz). However, the color signals do interfere with the luminance signal in many small ways. The most common readily visible effect is a color moiré pattern that appears on fine repetitive patterns in the picture, such as striped shirts and checked jackets. Many other transient effects can mar the image with color noise or impair fine luminance detail, but the result is an acceptable color picture. In addition, NTSC made a slight adjustment to the field rate, sending 59.94 fields per second rather than exactly 60, to prevent an interference problem with the audio signal.

The superimposed color signal does cause some interference, but it creates an acceptable color picture

Component Color Signals

Most professional color television equipment uses a component color signal system in which the chrominance signal is kept completely separate from the luminance signal. All professional video recorders store the video in component form. The signals are combined in a composite signal only at the transmitter. To reproduce the color image, the TV must separate the signals. For the first 25 years, color televisions used simple filters for separation. The widespread introduction of comb filters in the early 1980s improved separation, but once combined as a composite signal for transmission, the chrominance and luminance signals cannot be fully separated again, which accounts in part for the reduced quality of broadcast television vs. the original source.

Television sets receive the composite signal and must try to separate the color from the monochrome

Consumer Video Connectors

Only high-quality consumer video equipment can handle separate component signals

In consumer video equipment, the newer S-video connector supports component signals, while the simpler RCA and cable television F-connector can handle only composite signals. Even with an S-video connector, only higher quality camcorders (Hi-8, S-VHS) produce component images. Although VHS video tape records chrominance and luminance separately, the bandwidth available on VHS video tape is too limited to take any real advantage of the separate components, too limited, in fact, to show some common interference between the chrominance and luminance signals. The much higher-bandwidth laserdisc produces a much better picture, even though it stores video in composite form. The newer DVD format separates the two signals; many players make them available in component form.

Satellite Television

Satellite dishes point to a fixed spot in the sky

Satellite television broadcasting works quite differently from terrestrial television broadcasting. Satellites are very expensive to launch, and the broadcast coverage may be as large as an entire hemisphere, but satellites more typically broadcast to a smaller area (*footprint*), such as a country or group of countries. Most satellites are located in a geostationary orbit, 22,225 miles above the equator. Orbiting the earth 24 hours, they appear to remain stationary in the sky, so the receiving antenna, the now familiar dish, points at a fixed spot in the sky. For coverage in high latitudes, a few satellites operate on inclined or polar orbits; the receiving dishes must track such satellites as they move. More elaborate dish receivers can be moved to point at different satellites.

Satellite spectrum bands have 16 times more bandwidth than radio and television combined

Television transmission satellites operate on internationally agreed-upon frequencies in two bands—the C band, 3.4 to 6.425 GHz, and the Ku band, 10.95 to 14.5 GHz. Taken together, that's 6.575 GHz, or 16 times the combined spectrum bandwidths of AM, FM, and VHF/UHF television, which add up to only 408 MHz. (Other frequencies in the C and Ku bands are used by point-to-point satellites that handle general telecommunications.) With such massive bandwidth capabilities and wide coverage, it's easy to see the attraction of satellite broadcasting, especially once you also factor in the digital transmission capabilities of the satellites.

Each satellite carries multiple transponders

Satellites must operate in specific locations around the equator. For a given frequency, satellites must be separated by 2 degrees. However, satellites operating on different frequencies or aimed at widely different areas can occupy locations closer to one another. For obvious reasons, the positions over landmasses are much more desirable than positions over oceans. The satellite slots for North American coverage are completely full. Each satellite carries from 10 to 48 transponders operating in the C or Ku band or both. Each transponder acts independently, receiving a signal from an earth station (**uplink**) at one frequency and sending that signal back to earth (**downlink**) on a different frequency.

Satellite Dish Size

Small dish sizes are effective for receiving the powerful Ku band transmissions

The size of the receiving dish depends on several factors. The power of the transponder is one important variable. The more powerful the transponder, the smaller the dish can be. In addition, the smaller the coverage area, the greater the effective power in that coverage area, allowing for a smaller dish. C band dishes are large, typically 6–10 feet in diameter, while Ku band dishes are only 1.5–4 feet in diameter. Many transponders in the Ku band transmit with much more power than C band transponders, enabling the most popular receiving dishes to be a mere 18 inches in diameter.

Larger dishes can have better signal-to-noise ratios

Dish size also controls the quality of the reception, for a number of reasons. The larger the dish, the better the **signal-to-noise ratio** for the received signal. Also, dishes located at the edges of coverage—such as in the northern corners of the continental United States—often must be larger to maintain signal quality. Finally, Ku band reception is susceptible to interference from heavy rainstorms; a larger antenna can usually overcome such disruptions.

C band transmissions, initially geared to network affiliates and cable systems, require 8-foot dishes

C band television signals were originally intended only for relaying programs—both television and radio—to feed network-affiliated stations or cable systems across the country. The television industry never expected anyone else to install the 8-foot diameter dishes necessary to eavesdrop on the signals from satellites transmitting in the C band. But many people did, in large part because all the satellite channels were originally available without subscription fees.

Most satellite signals today are scrambled; subscriptions can be expensive

Today most of the programs sent via satellite are scrambled, with subscription fees similar to or even higher than cable TV. Nevertheless, people living in areas not served by either broadcast or cable TV or wanting more programming choices continue to use satellite dishes. In fact, some surveys show that satellite television systems are drawing as many as a million people a year away from cable. The main reasons are that satellite television can offer higher quality audio and video, plus significantly broader programming choices.

Satellite Bandwidth: Analog

Analog television transmissions use FM modulation

A transponder aboard a satellite can broadcast TV in either analog or digital form. The analog TV is sent in FM modulation rather than the AM modulation of terrestrial TV. FM modulation eliminates interference with adjacent satellites (remember that FM receivers lock onto the transmitted signal), but does use more bandwidth than an AM TV signal. An analog transponder in the C band typically has a spectrum bandwidth of 36 MHz, with the entire bandwidth being given to the TV signal. Some C band transponders send a half-bandwidth (18 MHz), lower-quality TV signal. In the Ku band, transponders use mainly 42-MHz bandwidths for TV broadcasting. A Ku transponder may also have 27-, 54-, 72-, and 108-MHz bandwidths; these typically handle data, but they can also be used for digital television signals.

Satellite Bandwidth: Digital

In digital mode, a 42-MHz bandwidth Ku band transponder can send a 23-Mbps data stream (after **forward error correction**), enough to support either four to six NTSC-quality television programs or one high-definition television program. The most popular Ku band services, such as the DSS broadcasts, send fully digital TV signals, the first consumer application of true digital television.

A 42-MHz Ku band transponder can support four to six regular programs or a single high-definition program

The attraction of switching from an analog to a digital signal for satellite television should be obvious. A single transponder generally supports a single channel of analog television, although some transponders can send two channels. But using digital signals enables either four to six channels of NTSC-quality television or one high-definition channel.

And that's even before you factor in the flexibility of being able to use the digital signal for other types of information. For instance, Hughes offers the DirecPC Internet service, which provides up to 400 Kbps of incoming bandwidth, received by a 21-inch dish that can also receive satellite television. Of course, because consumers have no way of transmitting back to the satellite, a standard telephone line and modem provide the outgoing bandwidth.

When the television signal is digital, the exact data rate or bandwidth of a particular signal can vary considerably. And the broadcaster can adjust the bandwidth for each signal within fairly wide limits—the tradeoff being that the narrower the bandwidth (the lower the data rate), the poorer the quality of the image. Digital compression makes this flexibility possible. (See the "Compression and Quality" section in Chapter 4 for more discussion of compression and other digital issues.)

Broadcasters can adjust the bandwidth of individual digital signals

Datacasting in the Television Channel

Just as with radio, there are a number of places within the standard broadcast television signal that can be used to carry non-video information. Over time, a variety of services have sprung up to exploit these bits of unused bandwidth.[1]

Vertical Blanking Interval

VBI originally carried no data

There are several places in the television signal that were originally wasted bandwidth but that can now be used for digital information. Earlier, I mentioned that television has 483 lines, but the usual specification is for 525 lines. This 525-line figure includes the black bar between fields, also known as the **vertical blanking interval** (VBI), which gives the receiver's electron gun time to jump from the bottom of the screen back to the top. Historically, it has been black because it usually contains no information and is only visible if your television has lost synchronization for the picture.

VBI can carry about 150 Kbps in the NABTS standard format

The VBI for each field contains 21 lines. The first 9 lines are used for timing and synchronizing the television picture. Line 19 is reserved for a ghost-canceling signal; a ghost-canceling TV measures the ghosts from this special signal and applies the inverse to the picture to remove ghosts. And line 21 of the odd-line field is reserved for closed captioning; the even-field line 21 is available for transmitting other information, sometimes going by the name of **extended data services** (EDS).

That leaves lines 10 through 18 and line 20 available for data. Each line can carry 288 bits (one **packet**) with a total bandwidth of 185 Kbps per second. After overhead and forward error correction, the net data bandwidth is about

---

1. You may be interested in reading an article about datacasting within analog television, published in the Corporation for Public Broadcasting's Info.p@ckets at *http://ready.cpb.org/library/infopackets/packet32.html.*

150 Kbps in an industry-standard format called the North American Basic Teletext Specification (**NABTS**).[2]

Several companies, including WavePhore and Intel, have come up with systems that use the VBI to distribute selected Web pages from the Internet, generally to special TV circuitry in a PC. Internet access via VBI is noninteractive and unidirectional; a standard telephone line with a modem would be required for anything that hasn't already been broadcast by the VBI Internet service provider.

Wavephore's WaveTop system stores the received Web pages on hard disk and serves them from there. This technique, called **caching**, can increase perceived speed beyond the **throughput** of the transmission medium; if you ask for a page that happens to be on the disk, the system can deliver it almost instantaneously. WaveTop delivers its data over the VBI of Public Broadcasting System programs; the software is built into every copy of Microsoft Windows 98.

Intel's Intercast uses virtually identical technology to transmit Web pages with information that's related to the current television show. And the totally unrelated StarSight system uses the VBI to deliver program schedules and descriptions to a device that then displays them on the TV screen. The StarSight device can also control a VCR, thus providing a better interface for programming the VCR to record shows.

North American Basic Teletext Specification

NABTS was originally developed for teletext, which can display information such as news and weather on the television screen. European televisions normally include teletext decoders, and broadcasters there routinely use teletext

NABTS is not expensive, but the data rate is too high to be recorded on a VCR

2. For additional information on NABTS, read the report on TV Data Broadcasting at *http://www.norpak.ca/article.htm*.

to deliver information ranging from airplane schedules to sports scores. In North America, NABTS has been a flop.

Because NABTS simply describes a way to send bits, it can carry any digital data, not just teletext. The NABTS data rate is too high to be captured by a VHS VCR, but the data is almost never directly related to the television program, so this does not normally matter. NABTS is not expensive; the equipment to insert the signal at the transmitter costs a few thousand dollars, and the receivers, based on readily available television tuners, sell for only $150–$350. In many countries, they are built into the television.

Closed Caption Data

Closed caption data, which supports 60 characters per second of uppercase and lowercase characters, is part of the program

While closed captioning is considered a type of datacasting, it's the only datacasting method that is actually part of the television program, unlike the other types of data sent in the broadcast television signal. Line 21 of the odd-line television field carries closed caption data; line 21 of the even-line field is available for transmitting other information, also known as extended data services (EDS). Closed captions were developed principally for the hearing impaired, but the service is also used for language learning and for saving transcripts of television broadcasts.

Closed captions support only 60 characters per second, a far lower bandwidth than the NABTS format. The lower bandwidth allows the caption information to be stored by standard VHS VCRs, which is convenient because the information is directly related to the television program on the tape. The captions are sent in upper and lower case with optional italics, although many early decoders displayed text only in upper case. The capturing of the closed caption text may be useful for archiving information about programs; an archive of captions from news programs could be a useful, searchable resource.

Data in the Picture Area

The FCC also permits data on line 22, even though it is the first line of picture. Line 22 data has been used to code programming information or to identify advertisements. (This deals with an old problem: you pay a station to run your commercial ten times. How do you know that the station actually did this?)

Proponents of using line 22 for data argue that the line is hidden by overscan. They make a similar case that the last line and a half of the picture or the left and right edges could serve for data as well. But there is a problem with taking over part of the picture and using it for data that looks like static on the screen: the overscan area has diminished as TV set design has improved. Besides, overscan is needed only for conventional TVs with standard CRTs (cathode ray tubes). Computers with video inputs and flat-screen television sets using LCD displays generally have no overscan, nor do they need it. The National Association of Broadcasters, the trade group representing television stations, opposed data on line 22 and has asked that no more of the picture area be turned over to data.

The edges of the picture become more crucial as displays eliminate the need for overscan

Figure 6-3 ***Modern flat-panel displays (left) can reveal the entire picture area, while CRT displays (right) cut off the overscan area.***

Modern flat-panel display

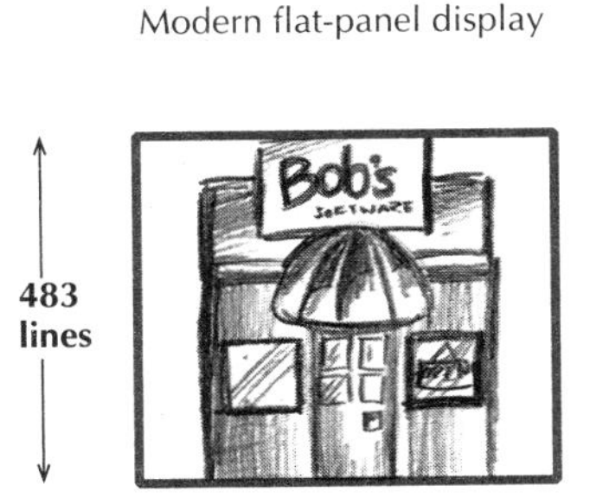

CRT display

Superimposing data over the picture creates interference on high-quality sets

Data can be put elsewhere in the video signal. There's space for a few bits during the horizontal **flyback** as the electron gun sweeps back across the screen; one company has proposed dropping an audio signal here, because a flyback frequency of 14,160 Hz (using lines 22 through 257) is enough to support audio.

And there are other possibilities. Data can be superimposed over the picture. The coding of data would make small parts of the picture lighter or darker. You'd see a noisier picture, but many people are used to noisy pictures already. The data rate depends on how much noisier a picture you are willing to put up with. NBC-TV experimented with such a system in 1992, but the data rate was only 260 bits per second. WavePhore has proposed a wider bandwidth system with 384 Kbps superimposed on the picture. These schemes use the high-frequency portion of the picture, producing visible interference—noise—on better television sets. Ironically, poorer quality television sets, with limited high-frequency response, display less interference.

Data in Video Sideband

The sideband is another place that can carry data

Digideck, a part of Datacast Partners, is testing a system to transmit data on the vestigial sideband of the television signal. By staying 1 MHz from the main carrier, the system minimizes interference with the picture, although observers say that interference is readily visible on high-quality television sets. There should be no interference with the adjacent lower channel because contiguous channels are not used in any given area. The signal would be transmitted over the air but would not be dependent on a television set or cable service. A card that goes in your computer would tune the signals; depending on your location, there could be multiple TV stations carrying the Digideck signal. The raw channel rate is 700 Kbps with about 25 percent of the bandwidth used for forward error correction, leaving an effective data rate of 525 Kbps.

Data Over the Entire Television Channel

Several groups have proposed using entire TV channels for data. This proposal is separate from the proposed **HDTV** broadcast channels that would carry video in digital form. The idea is to put data on unused channels, particularly away from urban centers, where many television channels are unused. This could support Internet services in a rural area that cannot have high-speed data service over phone lines because the wire runs are too long and the population density is too low for cable television. The return path—from the customer back to the Internet—would be done by low-speed telephone modem. Such usage of television channels would compete with other non-television uses that have started up in the past decade; in some areas, many of the higher UHF channels—first channels 70 through 83 and then channels 60 through 69—have already been reassigned for non-television services.

Unused channels can support data services

A company called En Technology has received approval from the FCC to extend the trial of its technology, called the Cybercast System, first tested in a television show called Internet Cafe. Cybercast encodes data in either the entire visible picture, replacing the picture with swirling dots, or only the VBI. At full speed, Cybercast claims top speeds of over 1 Mbps; when using just the VBI, speeds are around 128 Kbps. Unlike the Intercast system mentioned previously, Cybercast does not rely on NABTS. Cybercast requires special receivers that cost about $150.

Cybercast can use the VBI or an entire channel for data

Datacasting Standards and Distribution

Of all the various datacasting methods over television signals, only the VBI, closed caption, and the related line 21/even field data are established and standardized so far. Other systems may not succeed unless receiving hardware is produced in sufficient volume to be reasonably priced. Data over television, like data over FM radio, has one major advantage: the cost of the bandwidth is very low because these methods use no new radio spectrum or cables. The

Using existing radio spectrum or cables, datacasting over television allows broadcasters to control content and license use of bandwidth data content is up to the broadcaster, who can transmit news, traffic reports, television listings, Web pages, weather maps—whatever—or license the bandwidth to another company. Aside from closed captions, the data content will be mostly unrelated to the television program, although advertisers on television shows may provide additional information: you could get addresses of local stores that carry the product, or perhaps the inventory of cars at a local dealer.

While not required for broadcasting data over television, cable can extend the coverage

None of the methods of broadcasting data over television depends on cable television for distribution. If you are in antenna range, you would receive the data over the air to your computer even if you have cable television. Cable can, however, extend the coverage of the datacast, although cable systems in nearby areas or across borders may choose to strip off the broadcast data and substitute data of their own. Indeed, datacast weather and traffic information may not be relevant to a faraway cable area. Or a cable service that does not charge for carrying the television channel may charge for carrying data. Cable systems can and do handle VBI data easily; some other systems would need testing.

Worldwide Television Standards

There are a variety of television standards in use around the world.[3] All current systems use interlaced images. The Characteristics of the Main World Television Standards table shows the resolution and field rate for each standard, plus the countries that use each standard. In this table and in this section, the customary line counts are used, so NTSC, for example, is 525 lines rather than the actual displayed 483 lines.

3. For further comparison of different television standards around the world, check out the Worldwide TV Standards Web page at *http://www.ee.surrey.ac.uk/Contrib/WorldTV/*.

Characteristics of the Main World Television Standards

Standard	*Resolution*	*Field Rate*
NTSC (National Television Standards Committee, from the United States)	525 lines	60 Hz
NTSC is used in Antigua, Bahamas, Barbados, Belize, Bermuda, Bolivia, Burma, Cambodia, Canada, Chile, Colombia, Costa Rica, Cuba, Dominican Republic, Ecuador, El Salvador, Greenland, Guam, Guatemala, Honduras, Jamaica, Japan, South Korea, Mexico, Netherlands Antilles, Nicaragua, Panama, Peru, Philippines, Puerto Rico, Saipan, Samoa, Surinam, Taiwan, Trinidad, Tobago, United States, Venezuela, Virgin Islands		
PAL (Phase Alternation Line, from Germany)	625 lines	50 Hz
PAL (together with some variants) is used in Afghanistan (Kabul), Algeria, Argentina (PAL-N), Australia, Austria, Bahrain, Bangladesh, Belgium, Brazil (PAL-M), Brunei, China, Cyprus, Denmark, Finland, Germany, Ghana, Gibraltar, Hong Kong, Iceland, India, Indonesia, Ireland, Israel, Italy, Jordan, Kenya, Korea (North), Kuwait, Liberia, Luxembourg, Malaysia, Malta, Monaco (PAL & SECAM), Netherlands, New Guinea, New Zealand, Nigeria, Norway, Oman, Pakistan, Paraguay, Portugal, Qatar, Saudi Arabia (PAL & SECAM), Sierra Leone, Singapore, South Africa, Spain, Sri Lanka, Sudan, Swaziland, Sweden, Switzerland, Tanzania, Thailand, Turkey, Uganda, United Arab Emirates, United Kingdom, Uruguay (PAL-N), Yemen (A.R.), Yugoslavia, Zambia, Zimbabwe		
SECAM (Séquentiel Coleur Avec Mémoire, from France)	625 lines	50 Hz
SECAM is used in Albania, Benin, Bulgaria, Congo, Czechoslovakia, Djibouti, Egypt, France, Gabon, East Germany (former), Greece, Guadeloupe, Guyana, Haiti, Hungary, Iran, Iraq, Ivory Coast, Lebanon, Libya, Madagascar, Martinique, Mauritius, Monaco (PAL & SECAM), Mongolia, Morocco, New Caledonia, Niger, Poland, Reunion, Romania, Russia, Saudi Arabia (PAL & SECAM), Senegal, Syria, Tahiti, Togo, Tunisia, USSR (former), Vietnam, Zaire		

European television, with 625 lines, produces a distinctly sharper image than the 525 lines of North American television. All other things being equal, the European 50-Hz field rate requires less bandwidth than 60-Hz television because there are fewer images every second. But 50 Hz produces far more severe flickering than the 60-Hz in North America.

The European 625-line image is sharper; at 50 Hz, it requires less bandwidth but flickers more

Flicker is more visible with brighter and larger images.[4] To some degree, the visibility of flicker depends on what you are used to seeing; Europeans often claim to see no flicker on their television sets. Some recent high-end European TV sets overcome the flicker problem by doubling the display rate, although the process can introduce its own visual artifacts. The 60- and 50-Hz field rates follow the power line frequency in the respective countries, although there is no particular reason why the field rate must do so.

Video Bandwidths

Higher video bandwidth allows more detail

Aside from the field rate, the available video bandwidth limits the detail in the picture; higher video bandwidth allows for more detail. Thus the 4.2-MHz video bandwidth in North America is not as good as the 5-MHz in most of Western Europe. Eastern Europe, where standards were set later than in Western Europe, increases the bandwidth to 6 MHz. On the other hand, Argentina transmits a 625-line/50-Hz picture in 4.2 MHz because channel assignments in much of South America follow North American practice (6 MHz total bandwidth).

PAL and SECAM Color

In the 1970s, improved video distribution methods compensated for problems with NTSC color, making it better than PAL and SECAM

Early problems with NTSC color, particularly for long-distance relays, led to several refinements of the compatible color coding method. **PAL** (phase alternation line) from Germany became the most common standard in Western Europe; the French **SECAM** (séquentiel coleur avec mémoire[5]) was adopted in France, Eastern Europe, and some parts of the Middle East. Although they were a net improvement over NTSC at the time they were developed, both PAL and SECAM actually degrade the image slightly when used with the

---

4. Modern desktop computers mostly use 75- Hz or higher refresh rates to eliminate flicker.

5. Sequential color with memory.

improved video distribution methods available starting in the 1970s. The last industrialized nation to adopt television, South Africa, tested all the color standards in the early 1970s and concluded that NTSC color was the best when judged on a 625-line picture. In the end, the South African Broadcasting Corporation adopted PAL for easier program interchange with European countries (SECAM came in third). This situation illustrates the hazards of adopting standards like PAL and SECAM that compromise on image quality to address what may turn out to be temporary problems.

Choosing Color Standards

Economic ties influence the adoption of standards

Outside of North America and Europe, countries chose television systems based mostly on economic ties. Japan, Taiwan, the Philippines, and Central America follow NTSC (American) standards. Most of the rest of Asia uses PAL color. The former British colonies in Africa use PAL, the former French use SECAM. South America is mixed, with countries like Argentina using 625-line PAL in a 6-MHz NTSC-style channel bandwidth. Brazil uses PAL but with 525 lines. Some decisions were partially driven by a desire to have a unique system, giving the local electronics industry a boost. Indeed, individual companies or consortia that practiced restrictive licensing once controlled all three-color systems.

Conversion for international broadcasts sometimes creates a fuzzy picture

For international broadcasts, programs made in a particular standard must be converted to another. Standards converters have improved dramatically over the past two decades, but you can still see some artifacts; long smooth motion, such as a camera following a soccer game, will show slight discontinuities. The problem lies mainly in converting between 25 and 30 frames per second; the difference in the number of lines and the color coding simply makes for a somewhat fuzzy picture, particularly when going from 525 lines to 625 lines.

The standards-setting process is a complex one, filled with pitfalls. I've mentioned how both PAL and SECAM degrade

Setting standards too early can weaken quality and increase costs

the image in trying to solve what is now a nonexistent problem. If the standards are set too early, broadcasting is weakened by poor quality. The 1939 provisional American standard of 441 lines was replaced by a standard of 525 lines in 1941; RCA modified 441-line TVs to 525 lines without charge. Britain set a 405-line system (50 Hz, just 3 MHz video bandwidth) in 1938 and lived with the low-quality images until the 1960s, when 625-line color television arrived. And standards can be set so differently that other countries do not follow, driving up the cost of equipment and programming. The original French TV system had 819 lines with 10-MHz video bandwidth in part because of a justification that French culture could not be adequately represented with fewer lines. The French eventually retired the monochrome-only 819-line system in favor of 625-line color.

Television System Compatibility

Multistandard VCRs can play back all three standards; television sets often need external tuners for country-specific compatibility

Will a television set designed for one country work in another? Sometimes. The basic monochrome system must be the same (525 lines/60 Hz or 625 lines/50 Hz). The channel frequencies must match up; a foreign set may tune some but not all channels. Many other details must also match, including the separation between the picture and audio carriers and the polarity of the synchronizing signals. If the picture portion works, the sound may be different, or the stereo sound may be different. No television set made today works everywhere. Multistandard sets often need an external tuner for use in a particular country. The videotape situation is simpler; many companies make VCRs and monitors that can play back all three common VHS tape formats (NTSC, PAL, SECAM).

High-Definition Television

All current television systems have many faults. The most obvious is a fuzzy picture, particularly on large-screen TVs. The Japanese television industry, led by the Japan Broadcasting Company (NHK, after the Japanese-language initials), started research in the late 1960s to develop analog high-definition television. The result, Hi-Vision, displays 1125 interlaced lines at a rate of 60 fields per second, with a wide-screen aspect ratio of 16:9 (screen width to screen height—the aspect ratio used for many movies) rather than standard television's more boxy 4:3. Portions of the 1984 Los Angeles Olympic Games were taped in Hi-Vision.

Analog high-definition television can require 27 MHz of bandwidth

Hi-Vision is purely analog and needs 27-MHz bandwidth to transmit. That's too much bandwidth for any practical terrestrial or satellite broadcast, so NHK developed an analog compressed form, called MUSE, which uses 7-MHz bandwidth. Although NHK currently broadcasts Hi-Vision 14 hours a day by satellite in Japan, Hi-Vision is already obsolete because it is analog. Hi-Vision did spur the development of a full set of high-definition production tools, however, including cameras, videotape recorders, and editing equipment.

A compressed version of analog HDTV requires only 7 MHz of bandwidth

The electronics industries in United States and Europe responded to Hi-Vision with a variety of half-baked proposals in the late 1980s. These competing proposals for analog high-definition TV had only one important characteristic: they did not come from Japan. Yet none of the American or European systems could be built without critical components from Japan. Then in 1990, General Instruments proposed a digital HDTV system. That proposal drove everyone to dump analog plans in favor of digital. Eventually the four main development groups in the United States joined together as the "Grand Alliance," the Advanced Television Systems Committee (**ATSC**), to develop a single digital television system. The ATSC completed its specification for the Advanced Television System in September 1995.

ATSC formed to develop a single digital high-definition television system

The ATSC proposal supports one high-definition channel or several NTSC channels in the same bandwidth

The ATSC proposal defines a flexible digital television system with a data rate of 19.3 Mbps within a standard VHF/UHF channel; 19.3 Mbps can support one high-definition channel or several NTSC-quality channels, usually with some leftover bandwidth that can be applied to any purpose, such as datacasting. On a cable system, where error correction is much simpler, a single 6-MHz TV channel will allow a data rate of 38 Mbps. The television images are highly compressed, which enables transmission of much more data than would otherwise be possible in that amount of bandwidth.[6]

Until prices for digital televisions come down, digital tuners will provide an affordable alternative

Receiving HDTV will require either a new digital tuner or a new television that supports the higher resolution and the wide aspect ratio. Digital tuners are expected to be relatively inexpensive—in the $150 range—and will enable standard televisions to receive digital programming with good transmission quality and digital sound enhancement, but without the added resolution or aspect ratio. The new digital-capable televisions will initially be extremely expensive ($4,500) due to start-up costs and low initial demand.

The FCC has required the networks to begin digital transmissions; analog broadcasts may cease by 2006

Digital television will probably spread more quickly than it would have on its own, thanks to an FCC mandate that requires each of the four networks to have an affiliate with digital television capabilities in the top 10 markets (reaching 30 percent of the population) by May 1999. By November 1999, the network affiliates must begin transmitting digital programming in 20 additional markets, raising availability to roughly 50 percent of the population. To enable these actions, the FCC allocated 6 MHz of spectrum to every broadcaster for digital programming. The FCC's rollout plan for digital television continues into the twenty-first century, with analog

---

6. For additional details about the ATSC proposal, visit the ATSC Web site at *http://www.atsc.org/*. An excellent history of HDTV is *Defining Vision: The Battle for the Future of Television* by Joel Brinkley (Harcourt Brace, 1997).

broadcasts tentatively scheduled to cease in the year 2006. These schedules are now widely seen as unworkable.

One confusion about digital television is that broadcasters can in fact choose to broadcast either HDTV or **SDTV** (standard-definition television), which is purely digital but offers somewhat higher picture quality than today's analog television. SDTV requires far less bandwidth than HDTV, and in fact a broadcaster could send four or five channels of SDTV in the bandwidth required for a single channel of HDTV. Because the information is digital, broadcasters can mix and match. For instance, an HDTV broadcast of a fast-moving sporting event might require full bandwidth, whereas an HDTV broadcast of a game show (which has less motion and thus needs to send less data) could leave enough bandwidth free to send another program in SDTV. Once again, you can see how the flexibility of digital signals increases the value of bandwidth.

Digital television does not have to be HDTV; the same bandwidth can support several SDTV channels

The improved resolution of HDTV offers another important advantage over existing television systems: better compatibility with computer-generated text and imagery. Computer hardware and software companies are now lobbying to make sure that the progressive-scan formats they prefer for computer screens will work with HDTV receivers.[7]

HDTV is more computer-compatible

Motion Pictures

Although motion pictures are not a broadcast medium, the importance of motion pictures to video makes this a good place to discuss some aspects of them.

Motion picture film is shot at 24 frames per second, which matches neither the 25-frames-per-second or the 30-frames-per-second television system. For television with a 50-Hz

7. Two not entirely disinterested sources of information on HDTV are at *http://www.mot-sps.com/ADC/markets/HDTV/QA.html* and *http://www.cemacity.org/cemacity/gazette/vision/digtlage.htm.*

Motion picture frame rates are slower than television broadcast rates; television systems must compensate

field rate and 25-Hz frame rate, the movie is simply run 4 percent faster and the sound is pitched a semitone higher. It's a simple solution; few people complain, although people with perfect pitch sometimes object to music in the films shown on television.

For television with a 60-Hz field rate and 30-Hz frame rate, the conversion uses fields to correct the speed in a technique known as *3–2 pulldown*: odd-numbered film frames are shown for three fields, and even-numbered frames are shown for two fields. The technique produces some visual "stuttering" artifacts that are visible on large projection TVs viewed at short distance.

But 24 frames per second is too low for really smooth motion to begin with, especially when the entire scene moves. Cinematographers select the speed of camera motion with care. Some specialty motion picture systems have filmed and projected movies at as high as 72 frames per second (the Showscan system), but have not caught on because of the cost of installing new theater projectors to show such films.

Motion pictures carry more data per frame than television broadcasts

From a bandwidth standpoint, a frame of 35mm motion picture film holds much more information than a frame of standard television video. It's simply a denser storage medium. Film works very well as a source material for high-definition television because it provides sufficient resolution to showcase the improved display quality of HDTV. Thus, all of today's movies are candidates for HDTV broadcast, as are about 80 percent of today's television shows, which are also shot on film, many in wide screen.

In Conclusion

Broadcast television is a tremendously important aspect of communication in today's society. Estimates put the number of television sets in the United States alone at 250,000,000. In this chapter, we looked at the details of the television broadcast signal, how the television picture is composed, and ways of using space within the standard video signal to transmit digital information. Television is going digital in ever-increasing ways, ranging from today's methods of datacasting in the VBI and digital satellite television to tomorrow's HDTV broadcasts.

Television is a one-to-many broadcast medium, which makes it excellent for transmitting large quantities of information from a single source to many people. However, as we saw when looking at some of the Internet datacasting systems, this one-to-many aspect of television makes it a poor solution for customizing information or providing for two-way, interactive communications. For that, let's turn our attention to the next chapter, where we'll look at point-to-point communications, focusing on another ubiquitous communications device, the telephone.

Chapter Seven

Point-to-Point Bandwidth

Broadcasts go from a single transmitter to thousands or millions of receivers; in contrast, point-to-point communications connect two specific locations. This chapter begins with the most familiar form, telephone calls, and covers a number of related devices, such as modems, faxes, and pagers. Telephone terminology is often confusing, with several different terms in wide use for the same concept or function. I will try to use the clearest terms but will give some alternate terms where appropriate.

Wireline Communications

Because this chapter deals principally with the telephone system, I will start with the traditional telephone service with which we have all grown up. The term wireline refers to the fact that this service transmits data over wires that connect one point to another.

Public Switched Telephone Network (PSTN)

Most of the following discussion about telephone systems deals with the public switched telephone network (PSTN). Here's an explanation of each of the terms that make up the abbreviation PSTN.

- **Public** means that anyone can use the system. Not all telephones are on the PSTN; some phones operate on closed telephone systems that are not accessible to the public, such as secure military phone systems or corporate data systems.
- **Switched** (also called circuit-switched) means that the system operates point to point, switching one telephone to another. Once connected, the entire **channel** is devoted to the conversation or data (aside from the beeps or clicks of a call waiting). This technique works well for voice, but typically wastes bandwidth for data.
- **Telephone** means precisely what it says, although it now must be thought of as including devices like modems and fax machines.
- **Network** means that all the telephones are accessible to each other.

Original connections were direct

Early telephone systems achieved point-to-point connections in a simple, direct way. When you placed a call, an operator took a pair of wires that was directly connected to your telephone and plugged it into a jack connected to the telephone you were calling, as shown in Figure 7-1. The connection could not have been more direct. Some local telephone systems still connect two telephones directly with a purely analog path, but today most phone calls take a more complex route, where the signal is digital for most of the distance.

The local loop is usually analog

A telephone on its own phone line is connected through a pair of copper wires (the local loop or subscriber loop) directly to the local telephone company's central office. A local loop is dedicated to a single phone customer except in party line situations, where several customers share a single local loop, mainly in rural areas. In North America, the local loop is usually the only analog part of a telephone call. Thus, in a typical phone call, there will be two analog local loops—one at each end of the call.

The changing paths of telephone connections. ***Figure 7-1***

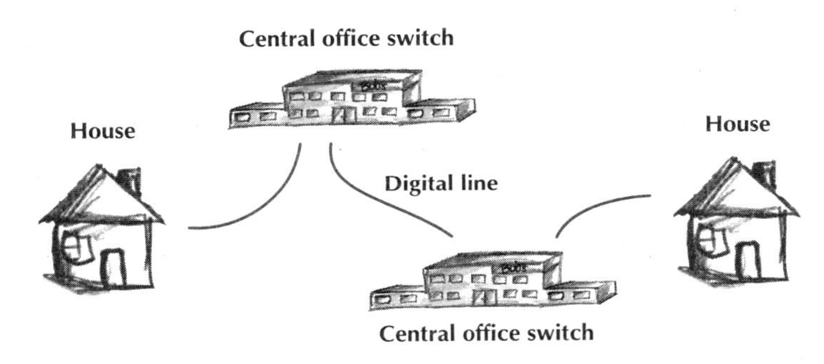

The local loop runs over telephone poles or underground and is the most variable part of the phone system. The copper wires may be old and in poor physical condition, causing noise on the line. Animals can cause problems; rats may gnaw on the insulation or birds or squirrels may block drain holes so that rain collects in distribution boxes and shorts out connections. Although phone companies are installing more wires all the time, some neighborhoods temporarily may not have enough copper pairs to meet the growing demand for modem, fax, and additional voice lines.

Local loops are vulnerable

I've pointed out that the local loop goes to the phone company's central office (CO). This name comes from a time when one central office might serve a town, but today the central offices are really local offices. The central office is sometimes simply called the switch (or telephone switch) because its main function is to switch calls; it's essentially an automated version of a human operator. Modern central offices are digital computers, so every phone call is converted

Except in rural areas, a central office (or switch) is a digital computer that converts calls to digital form and sends them on to a 64-Kbps digital data stream for switching, even for purely local calls connected to another local loop. Many older central offices, particularly in rural areas, remain purely analog. A single central office serves anywhere from a few hundred or a few thousand local loops in rural areas to over 100,000 local loops in city centers. The modular digital switching equipment is expanded as required.

Switches function in a hierarchy: low level for local calls, higher levels for longer distances

The roughly 25,000 central offices in the United States are class 5 switches, the lowest of five levels of telephone company switches. About 20 local switches feed into the higher class 4 switches, which digitize the signal from analog central offices, if necessary. A dozen class 4 switches feed into a class 3 switch, and so on. The farther the call goes, the higher it ascends in the switch hierarchy. Calls go up the hierarchy only as far as needed; a purely local call does not go higher than the nearest central office. The highest levels handle long-haul long-distance calls. The switch hierarchy was straightforward when AT&T was the dominant local and only long-distance telephone company, because most class 3 and all class 1 and 2 switches belonged to AT&T. Today, the switch hierarchy has become somewhat muddled after AT&T's breakup and because many companies have entered the telephone market at different levels. Yet the basic idea of a switch hierarchy remains.

Fiber-optic, coaxial, and copper pair wires carry the data between switches

Between the phone company switches, telephone calls travel in digital form on trunk lines. Unlike a local loop, a trunk line is always shared by many calls—hundreds or thousands of calls. Trunk lines come in many different forms. The major trunk lines are now mostly fiber-optic cable, but coaxial cables remain in wide use and short trunks with modest traffic may be copper pairs. Thanks to its vastly higher bandwidth, fiber-optic cable has now largely replaced the cross-country microwave relays that once served as the main long-distance trunk lines.

Telephone Bandwidth

As in any communications channel that goes through many steps, telephone bandwidth is limited by the slowest step. The limiting step may be a physical link, such as the wire, or a communications protocol, such as how the central office handles the phone line.

A copper wire local loop can handle several megabits per second

Contrary to widely held belief, the ubiquitous copper wire local loop that runs into every home and office is not a major bandwidth limitation today for most phone subscribers. A copper wire local loop can handle up to several megabits per second, a far higher bandwidth than it achieves or requires in its common use as a voice-grade phone line. If you are within two miles of the central office, the copper wire can carry 6 Mbps, enough bandwidth for digital (but not high-definition) television; about half of all phone subscribers live within a two-mile radius of a central office. If you are within three miles (as are nearly 80 percent of phone subscribers), the copper wire can surpass 1.5 Mbps, the rate of a high-speed T1 connection.

New services will be required for high throughput

There are, however, many caveats. The wire and whatever equipment it passes through must be in good condition, which, as I noted above, is often not the case, particularly in older or rural neighborhoods. These megabit-per-second **throughputs** will be available only with new kinds of equipment and tariffs. The distance limits are not exact; each local loop will have to be tested for performance, and throughputs may decrease with distance. In time, the installation of repeaters can give high bandwidth to subscribers beyond the normal distance limits.

The bandwidth limits of POTS lines restrict standard modems

The central office interface sets the bandwidth limits for ordinary phone lines. "Plain old telephone service" (**POTS**) subjects calls to an audio filter that limits the analog bandwidth to 4 kHz. And that limitation restricts modern modems, which in essence convert digital signals to sound and back again. The early 300-bps modem signals simply used tones of different frequency to code the data; modern modem signals use more complex modulations to achieve higher speeds. But there are theoretical and practical limits.

Modem throughput does not achieve theoretical limits

The central office interface digitizes the incoming analog signal of a POTS call into a 56-Kbps data stream and adds 8 Kbps of control information for a 64-Kbps total data stream that goes across telephone trunk lines. But although that 56 Kbps is a theoretical maximum, inevitable losses in converting analog to digital signals mean that modem throughput cannot normally reach 56 Kbps when an analog local loop is involved.

Actual throughput depends on the quality of the local loop

The potential modem throughput depends on the modulation technique; the limit has moved in recent years from 14.4 Kbps to 28.8 Kbps to 33.6 Kbps and beyond (see the Modem Standards and Throughputs table below). Modems at both ends of the connection must support a mutually acceptable throughput; a 33.6-Kbps modem connected to a 14.4-Kbps modem will "fall back" to 14.4 Kbps. The actual throughput achieved depends on the quality of the specific local loop;

Modem Standards and Throughputs

Throughput	***ITU Standard***
300 bps	V.21
1200 bps	V.22
2400 bps	V.22 bis
9600 bps	V.32 (also supports 4800 bps)
14.4 Kbps	V.32 bis (also supports 12 Kbps, 9600 bps, and 7200 bps)
28.8 Kbps	V.34 (also supports 26.4 Kbps, 24 Kbps, 21.6 Kbps, 19.2 Kbps, 16.8 Kbps, 14.4 Kbps, 12 Kbps, 9600 bps, 7200 bps, 4800 bps, and 2400 bps)
33.6 Kbps	V.34 bis (also supports 31.2 Kbps, 28.8 Kbps, 26.4 Kbps, 24 Kbps, 21.6 Kbps, 19.2 Kbps, 16.8 Kbps, 14.4 Kbps, 12 Kbps, 9600 bps, 7200 bps, 4800 bps, and 2400 bps)
56 Kbps	V.90 (older versions, not entirely compatible, were called K56flex and X2)

often a nominally 33.6-Kbps modem will operate at only 24 or 21.6 Kbps over a particular connection.

Because the conversion between analog and digital signals takes such a high toll on bandwidth, eliminating the conversions can improve bandwidth. As you will see in the following discussion of **ISDN**, a true digital line removes both conversions. As it turns out, eliminating just one of the local loops can also help.

Analog local loop conversions slow the throughput

New modem protocols developed by Lucent Technologies, Rockwell, and 3Com under the "56K" rubric do precisely this. You connect with your modem on your analog local loop, and the central office turns the signal into digital form; the signal then remains digital all the way to an Internet service provider or other online service and never undergoes another analog conversion. The key is that the service provider's modems must have a digital connection to the telephone company; smaller ISPs that have only standard analog connections for their modems must upgrade.

New modem protocols eliminate the need for conversions by the receiver

In these protocols, the downstream connection to you from the online service runs at a theoretical maximum of 56 Kbps; the protocols are asymmetric, with the upstream connection from you to the online service running at a maximum of 33.6 Kbps. An asymmetric protocol works fine for functions like Web browsing, where the online service sends far more information to you (Web pages) than you send back (mouse clicks and some keyboarding). But an asymmetric protocol does not work as well for functions like videoconferencing.

Asymmetric protocols are good for Web browsing, but not for videoconferencing

Lucent and Rockwell combined their protocols into the K56flex protocol, and the 3Com protocol is called X2. The two are mutually exclusive, which initially forced Internet service providers to decide between supporting one or the other, or to pay twice as much for equipment to support both. In December 1997, a compromise was reached between the two protocols, and the compromise standard under the

Once proprietary, protocols are now standardized

name V.90 is expected to be ratified officially by the **International Telecommunication Union** (ITU), the governing body of telecommunications standards, toward the end of 1998. V.90 modems are already on the market.

But "56K" modems are not really up to speed

These "56K" modems have proven useful but limited. Their asymmetric nature ensures that the outgoing throughput is never higher than 33.6 Kbps. Incoming throughputs never reach 56 Kbps either, in part because of FCC-mandated regulations on transmission power levels that limit the theoretical maximum to 53 Kbps and in part because high throughput requires a very clean local loop. And because these protocols rely on avoiding one analog-to-digital conversion, the modems cannot go beyond 33.6 Kbps in either direction when two individuals call each other over analog local loops. Admittedly, person-to-person modem communication isn't a major aspect of telecommunications today, but some businesses maintain dial-up connections for traveling employees.

It's possible to bond two lines, doubling connection speed

One way to get faster bandwidth combines two phone lines to behave as one virtual line. Two "56K" connections should yield 112 Kbps, for example. Until recently, bonding on analog modems has rarely been done; there are no standards for such bonding today. Two companies, Boca Research and Diamond Multimedia, offer products that bond a pair of "56K" modems,[1] breathlessly promising "up to 112 Kbps!" Both enable you to use call waiting to answer calls on one line while the other remains connected via a modem, and Diamond's technology can be set to add the second line automatically when traffic becomes heavy.

---

1. To read more about the Boca Research and Diamond Multimedia bonding technologies, check out the companies' Web sites at *http://www.bocaresearch.com/docs/mddl56i.htm* and *http://www.diamondmm.com/shotgun/*.

Of course, should these products become widespread, some questions should be raised. First, what throughputs are realistic? Existing "56K" modems never reach 56 Kbps on incoming traffic and never do better than 33.6 Kbps on outgoing traffic. It's safe to say that claims of 112 Kbps are misleading at best. Second, such products require two phone lines, so you can double the amount of money you pay the phone company for your phone lines each month. In addition, although you may be able to connect each line to the same Internet account for now, Internet service providers will undoubtedly react by restricting that access unless you pay an additional fee. Once you add up these extra fees and add them to the cost of a second modem, will the overall cost be lower than better alternatives, such as the fully digital ISDN?

Costs would also double; other alternatives may be better

ISDN

ISDN (Integrated Services Digital Network) has been coming for several decades, as part of an overall plan to move the telephone system from the analog world to an entirely digital network. In short, ISDN makes the local loop digital. ISDN runs on the same pair of wires that the analog POTS service uses; the phone company simply makes a change at the central office so the ISDN line bypasses the analog-to-digital conversion necessary for POTS service.

ISDN uses the same local loop but eliminates the need for central office conversion

From a technical standpoint, ISDN can be simpler and cheaper to support than a POTS line because it is already digital, although existing equipment sometimes must be replaced to support ISDN, and existing wires may not be sufficiently clean either. From a practical marketing standpoint, phone companies argue they should charge more for ISDN because it offers additional features and because ISDN is more expensive to service and support—partly because it is new and partly because of the lack of complete standardization for ISDN hardware.

ISDN BRI has two bearer channels and one data control channel

The standard ISDN connection—ISDN BRI (Basic Rate Interface)—is actually three digital channels on a pair of wires, as shown in Figure 7-2. BRI has two B (bearer) channels of 64 Kbps each, plus a D (delta) channel of 16 Kbps, often described as "2B+D." Some phone company equipment restricts the B channels to 56 Kbps, which causes some confusion because it means that ISDN throughputs can be 56 Kbps, 64 Kbps, 112 Kbps, or 128 Kbps, depending on the location and number of channels used. All the channels are digital.

Figure 7-2 ***ISDN bandwidth includes two 64-Kbps bearer channels and a 16-Kbps data channel.***

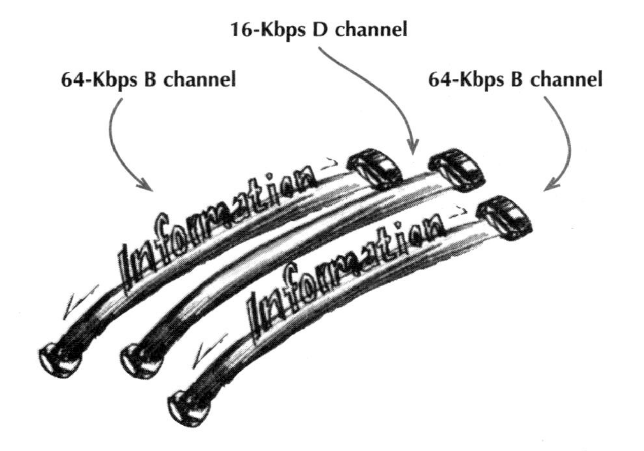

The D channel is intended to support calling information; for example, dialing instructions are sent separately in the D channel, in contrast to the POTS practice of dialing in band, which you can hear during a call. An ISDN call takes only a few seconds to set up, provided the call is to another ISDN phone. The D channel can also be used for sending digital data, although this is mostly restricted to very small amounts of data, such as magnetic stripe data on the back of a credit card. Such D channel usage does not interfere with voice or data on the B channels.

The B channels are modular building blocks. If you want more bandwidth than a single 64-Kbps B channel, you can bond two B channels together for a 128-Kbps channel, provided that the ISDN hardware permits such bonding. For even higher bandwidth, more B channels can be combined, either from multiple ISDN lines or higher capacity such as T1.

B channels can be bonded

If the ISDN terminal hardware is suitably equipped, each ISDN B channel can behave like a conventional phone line; you simply plug in ordinary telephones just like a POTS line. So ISDN's two B channels can serve to double the number of phone lines when the number of copper pairs is short. POTS does have one important advantage over ISDN for voice service: ISDN terminal equipment depends on external power. A POTS line is typically operational even when AC power has failed, which is handy for calling the electric company to report power outages. Thus very few locations have only ISDN service installed.

Analog telephone equipment works on ISDN lines, but it requires external power

Because ISDN line characteristics are more precisely defined than POTS, an ISDN line can carry higher quality audio than POTS. Many radio programs use ISDN lines for voice relays whenever available; a whole category of professional audio equipment is designed around ISDN. Similarly, some office telephone equipment is designed around ISDN to take advantage of audio quality and fast D channel switching features not available on POTS lines.

ISDN line can carry higher quality audio than POTS lines

One useful feature of ISDN connections is that they can switch quickly between voice and data use. For instance, assume you have a two-channel 128-Kbps ISDN connection to your Internet service provider. If a call comes in while you're connected, your ISDN hardware can be configured to drop the Internet connection down to a single 64-Kbps channel and let you take the voice call. Then, when you hang up from the voice call, the ISDN hardware brings the second 64-Kbps channel back into use for the Internet connection.

ISDN connections can automatically switch between voice and data

Packet-Switched Data Communications

Both POTS and ISDN were designed for continuous voice communications, not bursty data transmissions. Although POTS phone lines are analog and ISDN phone lines are digital, both were designed for voice conversations rather than data, even though ISDN is mostly used for data today. This means that both have serious disadvantages for data. Both are circuit-switched connections designed to stay in place as long as you are expecting to send and receive data, rather than when you are actually sending and receiving data.

Most data connections are inherently "bursty": you send or receive some e-mail or Web page information in a burst. Then you may spend a lot of time reading the information before replying to the mail or going on to another Web page. Yet during all the time you are reading and thinking, the POTS/ISDN connection remains in place, tying up a line through the central office to an Internet hookup even though there's no data traveling on the line.

Packet-switching allows multiple users to share the communications channel and maintain a continuous connection

This inherent flaw—the "misuse" of a continuous voice channel to carry bursty data—has enormous consequences for everyone who uses the phone system. Bursty data should properly be carried on a packet-switched communications channel, which, as discussed in Chapter 3, is shared by multiple users.

With multiple users, the bursts of data and quiet periods average out statistically. The typical local or wide area network uses a **packet** protocol, as do many wireless protocols. Besides using a communications channel more efficiently for digital data, packet connections typically let you connect to the network continuously. With an "always-on" connection, you send and receive information whenever the occasion arises; you don't have to dial into a modem system first. You don't have to check to see if any e-mail messages are waiting; your computer will always show you.

For one important application, circuit-switched connections have an advantage over packet-switched: because you get the full bandwidth of a circuit-switched connection, information like voice and video that requires a steady, continuous stream of data can travel without disruption, at least as far as the specific connection goes. Complex data networks such as the Internet often mix circuit-switched with packet-switched segments, so the actual available bandwidth is complex and changing. Advanced packet-switched communications channels, including proposals for special Internet protocols, can set aside guaranteed bandwidth for specific functions such as voice and video.

Advanced packet-switched channels can guarantee bandwidth for specific functions such as video

For a given bandwidth of digital data, packet-switched connections are usually cheaper than circuit-switched, but cost, like other distinctions between packet-switched and circuit-switched connections, depends on many factors, and no absolute generalizations are possible. For example, in some installations, multiple users can share a circuit-switched connection, in which case the connection takes on some of the attributes of a packet connection. Or the circuit-switched connection can be established at regular intervals. You can set up a circuit-switched data network to send and receive on demand—if you are willing to stay connected continuously.

Some circuit-switched connections can behave in packet-switched ways

As you saw with ISDN, replacing the analog local loop with a digital local loop improves bandwidth. The same copper wires are still used, except that they carry digital signals. But ISDN still suffers from being a circuit-switched network, as do the older digital "switched 56" and "switched T1" services.

A digital local loop can help

Frame relay and leased lines There are two main types of packet-switched services that offer digital service at throughputs generally ranging from 56 Kbps to 1.5 Mbps (T1). First comes frame relay, which is an economical solution to the need for bursty data. Second, if you need a guaranteed amount of bandwidth, you can get a point-to-point leased line. The difference between these services, like so many other bandwidth-related issues, is one of bandwidth vs. cost.

The bandwidth on a leased line is all yours

It's easy to visualize a point-to-point leased line: A wire essentially travels from your home or office to the phone company, and then out to your Internet service provider. You don't share that connection with anyone and thus are guaranteed the full bandwidth of the connection (which is determined by how the phone company configures its equipment—throughputs can range from 56 Kbps to 1.5 Mbps).

In frame relays, PVCs give the appearance of leased lines

Frame relay connections, depicted in Figure 7-3, are somewhat more complicated. Think of the connection traveling from your location to a "cloud" maintained by the phone company. All frame relay connections terminate in that cloud. Then, other connections lead from the cloud to your Internet service provider. So, although you don't have wires going directly from you to your Internet service provider, you have a virtual connection, called a **Permanent Virtual Circuit** (PVC), that makes it seem as though you have that leased line.

Figure 7-3 ***The Frame Relay Cloud.***

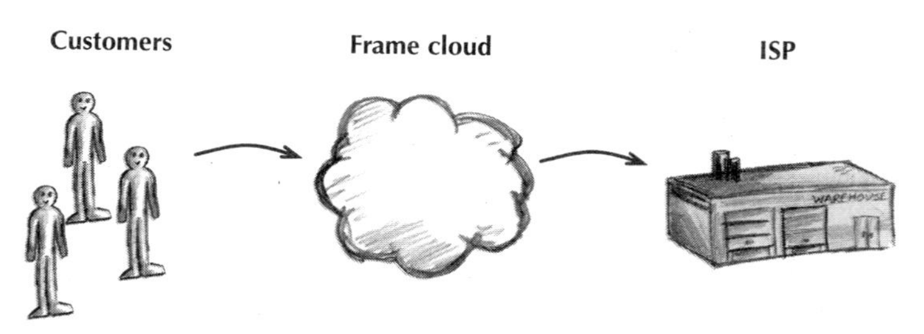

Frame relay is cheaper than leased lines...

The reason for the cloud is that most packet-switched connections aren't fully used most of the time, meaning that there's extra bandwidth available. With frame relay, the phone company can take advantage of that extra bandwidth and sell more connections without adding any additional bandwidth. Hence, the price for frame relay connections is lower than for point-to-point leased lines.

There is, of course, a catch. Frame relay connections have what is called a Committed Information Rate (CIR), which is the rate that the phone company guarantees. For instance, a 56-Kbps frame relay connection might have a CIR of 48 Kbps. However, if more bandwidth is required for a data burst, and if the entire frame relay network (the cloud) has more bandwidth available, the extra bandwidth is provided up to the nominal rate—56 Kbps in this case. It's even possible to have a CIR of zero, at which point you'd rely entirely on whatever spare bandwidth is available.

...but the information rate guaranteed by the phone company may not be high

Digital Subscriber Line service An up-and-coming technology in this field is DSL, which stands for **Digital Subscriber Line**. DSL comes in a number of variants, each with a different initial, so it is sometimes known generically as xDSL. The most common is ADSL (**Asymmetric Digital Subscriber Line**), which provides much more throughput coming in than going out, similar to the 56K modems. Incoming ADSL throughputs can range as high as 1.5 Mbps; outgoing throughputs can be as high as 512 Kbps. The throughputs in both directions depend on a number of factors, including your distance from the central office, the quality of your local loop, and of course, how much you're willing to spend.

Throughput on DSL varies with distance, equipment quality, and price

The leading contender for the consumer DSL market appears to be a technology called Universal ADSL, or DSL Lite. Based on technology from Aware, Inc., DSL Lite provides an asymmetric connection at the previously mentioned throughputs of up to 1.5 Mbps incoming and up to 512 Kbps outgoing. DSL Lite is "rate-adaptive," meaning that throughputs can vary dynamically to adjust to local conditions. DSL Lite's rate-adaptive nature enables it to work well in consumer situations, since distance merely reduces bandwidth rather than creating an either/or situation where the connection either works at a high throughput or doesn't work at all.

Rate-adaptive DSL Lite is favored for the consumer market

An important fact about DSL Lite technology is that it runs over the copper pairs that exist in houses now, requiring

DSL Lite, running over existing copper wires, supports a continuous data connection at the same time as standard voice service

only a special modem and support from the telephone company. Over that standard pair of copper wires, DSL Lite provides simultaneous voice service and constant access to the Internet—no need to dial in to send or receive e-mail messages. This "always-on" aspect of DSL is probably one of the most important parts of the technology for interactive communications, because, for the first time, your computer can "ring" to alert you to an incoming connection.

By avoiding the 0 to 4-kHz voice portion of the signal, DSL Lite eliminates the need for a splitter

Previous DSL technologies required a "splitter" to achieve this; the splitter broke apart the voice call from the data but it was expensive to install, because it required a service visit (the industry term is "truck roll") from the phone company. Because DSL Lite can use 500 kHz of spectrum, the protocol merely avoids using the 0 to 4-kHz voice portion of the signal for data, eliminating the need for a splitter and the associated service call. When the signal reaches the central office, special DSL hardware separates the voice signal from the data signal and feeds them into the phone company's existing voice and data networks, respectively.

Maximizing use of existing infrastructure could keep costs affordable

Aside from the benefit to the consumer, this model has great advantages for the telephone companies in terms of cost. As you've seen, it's wasteful to send data over a circuit-switched network like the voice network. By utilizing the voice network only for voice calls and sending the data over a packet-switched network, the telephone companies make much more efficient use of their infrastructures, which in turn enables them to charge much lower prices than would otherwise be expected. In addition, the telephone companies have had to install additional copper pairs and switch capacity to provide people with the second phone lines for the modems and faxes that have become so popular. Combining voice and data on a single pair of wires might enable the phone companies to recover some of the copper that's currently installed and use it for new customers, minimizing the need to install more. Pricing for DSL Lite service is predicted to be in the $40 per month range. However, that may be for the slowest throughputs and may or may not include Internet service as well as the connection.

Some people are concerned that DSL Lite will solve only one part of the overall bandwidth problem by providing much more bandwidth from the Internet service provider to the individual. As you will see in the next chapter, however, there are many other potential bottlenecks in Internet traffic. For instance, if an Internet service provider can afford to sell 10 times more bandwidth than it buys when most users connect via modem, how will that ratio change when users have multiplied their bandwidth capabilities 3, 5, 10, or even 30 times? Unless the Internet service provider charges far more for people to use faster connections or bases rates on how much data is actually transferred, it won't be able to afford to buy a sufficient amount of bandwidth out to the Internet. It would be a bit like hooking two fire hoses together with a standard garden hose and wondering why so little water is coming out.

Increasing bandwidth between consumers and their Internet providers does not solve all problems, and it may create some new dilemmas

Consider the following numbers, which may no longer be accurate by the time you read them, but they're useful as a guide. A full 45-Mbps T3 connection to the Internet costs an ISP $65,000 per month. That equals roughly $1,450 per month per Mbps. Currently, U S West, one of the regional Bell operating companies, charges $40 per month for the lowest capacity (192-Kbps) ADSL line (note that this isn't DSL Lite, which hasn't yet been priced). For 192 Kbps of bandwidth, straight to the Internet backbone, an ISP would have to pay about $275 per month.[2] So an ISP has to make at least $275 from a customer to pay for that bandwidth. In fact, the ISP has to make more than that, about double, or $550, to account for overhead and profit margin.

To pay for a 192-Kbps ADSL connection to the Internet backbone, an ISP has to make a few hundred dollars per month per customer

2. 1 Mbps is about 5.3 times as much bandwidth as 192 Kbps. So, if we divide $1,450 by 5.3, we get roughly $275.

Charging customers only $20 per month means an ISP must oversell the connection; the result is haphazard service

ISPs currently charge about $20 per month for a modem connection that can't go faster than 56 Kbps. If we assume that consumers aren't willing to pay more than that $20 per month, no matter what the speed (which equals a total bill of $60 per month, once you add the line charge), you can divide our $550 by $20, which is about 28. Thus, to be able to afford to sell access at $20 per month, an ISP would have to sell that 192-Kbps connection 28 times. If you consider that ISPs generally aim for an oversell ratio (how much more bandwidth they sell than they buy) of 5:1 up to 10:1, you can see that a ratio of 28:1 might result in haphazard service. And that's at the lowest capacity ADSL connection currently available from U S West; increase the throughput of the connection and you increase the necessary oversell ratio.

Small ISPs may not survive

Of course, if the telephone company acts as the ISP, that eliminates some of the overhead and brings the costs and thus the oversell ratio down. That will hurt the smaller ISPs more than the larger ones, since it will be harder for them to negotiate terms. One of the legacies of ADSL may be to speed up the consolidation process that's forcing the elimination (mostly by acquisition and attrition) of small ISPs.

Business and pricing models will change

In short, although ADSL sounds like a technology to watch, it's almost certain to cause some significant changes in existing ISP and telephone company business models. That's not necessarily bad, but nothing else has been doing much to force ISPs toward tiered pricing (where you pay only for the data you use) and eliminating the "wasteful" use of the circuit-switched network for data.

Cellular Communications

Cellular phone calling has been the fastest growing portion of the telephone system in recent years. Earlier radio-telephone systems were never popular because only a few frequencies were available, with a single transmitter and receiver for an entire metropolitan area; thus no more than a handful of calls could be placed at the same time. Cellular phones solve the congestion imposed by limited radio frequencies by reusing the same frequencies over and over so that many thousands of calls can be placed at the same time.

Cellular calls reuse available frequencies

Analog Cellular Systems

In North America, the most common cellular system is the analog **Advanced Mobile Phone System** (AMPS). AMPS assigns a pair of 30-kHz FM analog channels to each call, one channel for the cell phone to transmit on, the other to receive on. The bandwidth of the phone call itself is 3 kHz, a bit less than a POTS phone call. At least one provider uses a narrow band version of AMPS that performs frequency compression, reducing the channel bandwidth to 15 kHz and doubling the number of available channels. Few people can hear the difference.

AMPS is an analog system using channel pairs

In cellular terminology, a wired POTS phone is a *wireline* phone. In each service area, the FCC has licensed two cellular service providers. One is the company that runs the local telephone system (the wireline carrier; originally one of the Bell operating companies) and the second is a non-wireline company (the largest is now AT&T Wireless, formerly McCaw Cellular). As you can see in the Transmit and Receive Frequencies for Band A and Band B table on the following page, two sets of frequency bands carry the calls. The second frequency assignment was added after the original specification to allow more channels.

Two sets of frequencies are used to accommodate cellular traffic

Transmit and Receive Frequencies for Band A and Band B

Band	*Cell Phone Transmit*	*Cell Phone Receive*
Band A	824–835, 845–846.5 MHz	869–880, 890–891.5 MHz
Band B	835–845, 846.5–849 MHz	880–890, 891.5–894 MHz

Band A is for the wireline carrier, band B for the non-wireline carrier. Because the channels are separated by frequency, the AMPS technology is sometimes described as frequency division multiple access (FDMA).

Most regions' cells are hexagonally shaped and include a base station with a 100-watt transmitter/receiver

With 30 kHz per channel, each cellular operator has 416 channel pairs available; one or two control channels in any cell are used to manage the calls—identifying cell phones, assigning the channel to use for a particular call—instead of carrying a phone call. To support thousands of calls simultaneously, each geographical region is divided in hexagonally shaped cells. Each cell contains a base station, a 100-watt transmitter and receiver, in some high place—a tower, building, or hilltop. The base stations in a region are connected to and controlled by the cellular provider's **Mobile Telephone Switching Office** (MTSO).

Phone range depends on transmitter power; remote areas may not be served

The hexagonal cell shape is idealized for flat terrain; the actual shape can vary, and to cover large cells in remote areas, the base station may use over 100 watts. At the frequencies used by cell phones—between 824 and 894 MHz—the signal travels line-of-sight. Just like a wireline phone system, cellular phone systems are expensive to build; cell sites are installed where justified by the number of users. Thus entire cities and settled areas between cities are served, but not thinly settled open country.

Despite this avoidance of open country, remote areas with the right demographics will often be served—ski slopes, for example, and areas near busy interstate highways. Because of local geography, coverage may be erratic or absent in canyons, deep valleys, and sometimes in localized areas within a city. The range of a cellular phone depends on its transmitter power; high power is not possible in the tiny phones that many users demand.

***Cell phone coverage for the entire country.*[3]** **Figure 7-4**

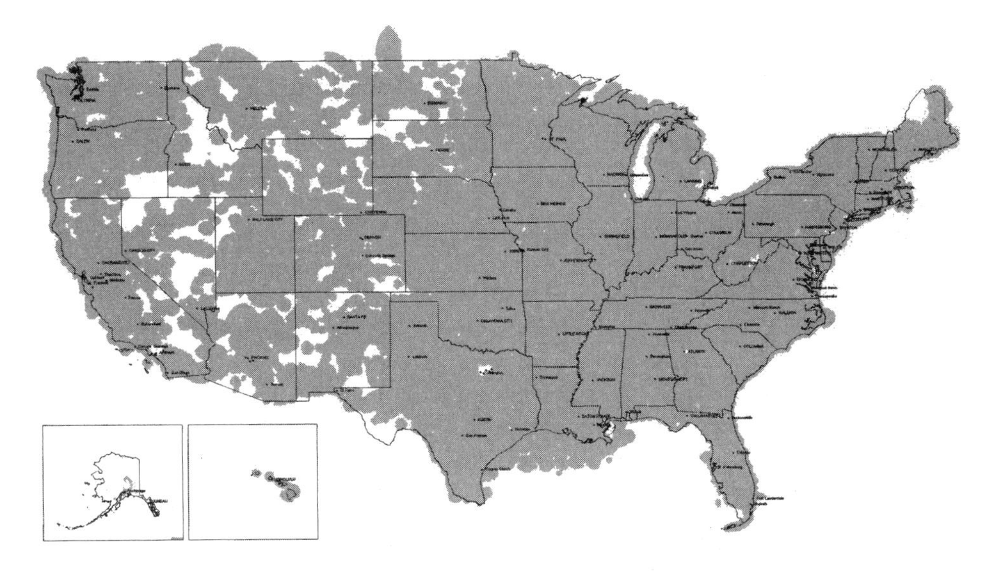

The MTSO assigns channel pairs and tracks location

The MTSO directs all the call activity.[4] When a cellular phone initiates a call, it sends a request on the control channel for a channel pair, which the MTSO assigns. If the cellular phone moves from one cell to another, the MTSO tracks the location through the adjacent base stations and hands off the call from one base station to another. The user will often hear a brief interruption during the handoff.

---

3. This image was based on a graphic seen at *http://www.fcc.gov/wtb/cellular/cel_cov.html.*

4. For another description of how cell phones operate, read the cell phone section on the *How Stuff Works* Web site at *http://www.bygpub.com/HowStuffWorks/cell-phone.htm.*

Figure 7-5 ***As you move in and out of a cell, your phone switches from one tower to the next.***

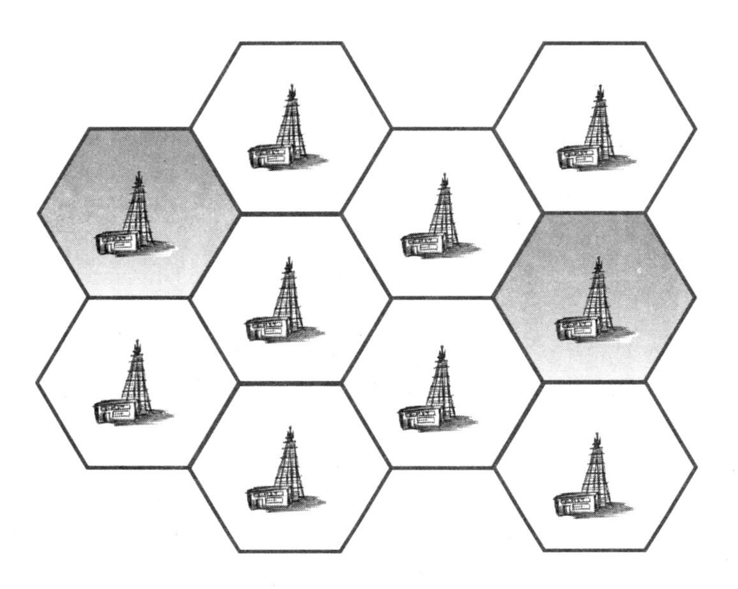

To minimize interference, the MTSO adjusts the phone's transmitter power

Because of **interference**, a specific frequency cannot be used by two adjacent cells, so only one-quarter of the channel pairs are available for a specific cell. Because a very strong signal could span two cells and cause interference, the MTSO adjusts the transmitter power of the cellular phone from 6.3 milliwatts to 0.6 watts for a handheld phone, or 4 watts for car phones. The much higher 100-watt power of the base station permits the cellular phone to have a very small antenna, while the large receiving antenna at the base station can pick up the low power signal from the cellular phone.

Cell calls placed from high altitudes may connect with multiple cells simultaneously

Cellular phone usage is prohibited on airliners because the transmitter inside the phone might interfere with equipment on board the aircraft. Quite aside from the safety issue, a cellular call placed from a high altitude tends to confuse the cellular system, "lighting up" many cell sites simultaneously. The same phenomenon may occur when calls are made from mountain peaks overlooking many cell sites.

The cellular design gives the system its capacity; as the calling traffic increases, the cellular provider can split cells into smaller and smaller units, reusing the frequencies over and over in smaller and smaller areas. Radio systems do not work perfectly, however, and a variety of problems complicate the process of dividing cells. Nevertheless, cellular phone systems have proven practical for voice conversations.

Cellular providers can split cells into smaller units and reuse the frequencies

At the beginning of 1997, there were about 80 million wireless phone users worldwide; about half were in the United States. The number of cellular phone subscribers grows at about 30 percent per year. When the AMPS cellular system was designed back in the 1960s, the planners predicted one million users in the United States by the year 2000, a prediction that many considered wildly optimistic at the time.

Cellular phone subscriptions grow 30 percent per year

Digital Cellular Systems

Several digital cellular telephone technologies increase the number of available channels over AMPS. Of course, the cellular provider could increase the effective number of channels by reducing the cell sizes, but that's more expensive and has limits. Increasing the number of channels can put off the day when the cells need to be reduced. Because of limited channel capacity, all cellular providers are phasing out the analog AMPS system, although the process will take years and many new AMPS phones are still being made and sold. Some cellular providers charge lower rates for digital service to encourage users to switch to the higher capacity service. Because many areas do not have digital service yet, some cellular phones are dual-mode designs, using digital if possible, analog when necessary.

Analog is slowly giving way to digital cellular; some cell phones accommodate both

The digital cellular systems work in the same basic manner as AMPS, except that the voice information is transmitted digitally. The MTSO assigns digital and analog channels as required to prevent interference, just as in an all-analog

Digital cellular supports twice as many channels, using the same frequencies as analog

system. In North America, the digital cellular systems work over the same frequencies as AMPS, allowing twice as many channels in the same frequency spectrum through very high compression of the voice signal; the effective audio bandwidth is lower than for even narrow band AMPS.

Voice coders for digital systems provide inferior voice quality

The key to the compression is a vocoder—a voice coder—built into the cellular telephone. Thus far, the vocoders for the digital systems produce sound quality distinctly inferior to AMPS. The vocoders work poorly because they work with so little bandwidth; if they were given a full AMPS channel bandwidth to operate in, they would have better sound quality.

PCS services include wireless phone, data, and paging

The new **Personal Communication Services** (PCS) in the United States use 120 MHz of radio spectrum in the 1850–1990 MHz range auctioned off by the FCC. The PCS services include wireless phone, data, and paging. All will operate as digital systems and will compete with the existing cellular systems for phone service. There are six separate PCS licensees in each market, which means that, including the two "traditional" cellular systems, up to eight companies may offer cellular phone service in a given city.

Digital Cellular Technologies

The cellular phone industry has adopted several incompatible digital technologies.

- **Time Division Multiple Access** (TDMA) is being installed by AT&T Wireless in North America.
- **Code Division Multiple Access** (CDMA) is being installed in the United States by most competing carriers, including the wireline carriers.
- **Groupe Special Mobile** (GSM) is a form of TDMA technology widely used around the world for digital cellular phones and employed in the United States by some PCS carriers.

TDMA splits each 30-kHz analog channel into three independent 8-Kbps segments, tripling the effective number of channels. The digital modulation in TDMA is for voice only and cannot handle data, either as a modem signal sent over the phone connection or as a direct data connection, although the protocol may be extended in the future for direct data connection.

TDMA triples the number of channels but is for voice only

CDMA uses spread spectrum to place 64 channels in 1.25 MHz of radio spectrum. All 64 channels use the same spectrum, with a frequency-hopping coding scheme that allows receivers to separate the channels. The unwanted channels act like noise; if all the channels are in use, the signal quality—the **signal-to-noise ratio**—of the desired channel degrades. As callers hang up, the signal quality improves. CDMA handles the cell-to-cell handoff more elegantly than AMPS or TDMA, making the new connection before breaking the old one, eliminating the signal dropout. Some PCS companies use CDMA variants.

CDMA can have high signal-to-noise ratios, but it eliminates signal dropout

GSM was originally developed by a group of European telephone companies. Unlike the incompatible AT&T TDMA protocol, the GSM format can handle digital data at 9600 bps for a stationary phone. For a voice call, the phone can move. GSM supports a paging mode that can transmit a 160-character message for display on the GSM telephone in the manner of a paging receiver. This feature, which combines digital features with cellular telephone, will undoubtedly spread to all digital cellular systems over time.

GSM can handle mobile voice and stationary data transmissions

The GSM technology can be applied at varying frequencies; at present there are three main types of GSM systems—see the Transmit and Receive Frequencies Used by Different GSM Variants table on the following page. Most European countries are switching over to GSM on the standard European cellular frequencies, replacing the six incompatible analog cellular systems that were installed earlier. A standard GSM phone works throughout Europe. A higher frequency variant

PCN is a high-frequency variant of GSM, operating at lower power

Transmit and Receive Frequencies Used by Different GSM Variants

GSM Variant	*Cell Phone Transmit*	*Cell Phone Receive*
Europe standard	890–915 MHz	935–960 MHz
Europe PCN	1710–1785 MHz	1805–1880 MHz
United States PCS	1850–1910 MHz	1930–1990 MHz

of GSM called Personal Communication Network (**PCN**) works at a higher frequency band and operates at lower power than standard GSM. PCN is starting to become widely available in Europe, Asia, and Oceania. PCN is similar to the **PCS** services in the United States, many of which will use a third variant of GSM.

Roaming is possible in areas with compatible protocols

If you use a cellular phone outside the area it was originally registered in, you are **roaming**. Roaming is generally possible in any area with a compatible cellular protocol, which eliminates most countries for people from the United States, although Australians, for instance, can roam with no trouble in Europe. Fees for roaming vary considerably; some new PCS systems have eliminated roaming fees altogether, at least as an initial promotion. No single telephone will work everywhere because no one cellular protocol operates worldwide. A multi-standard set could be built, but it would inevitably be larger and more expensive than a single-protocol cellular phone.

In America, cellular costs are high and cell users pay for calls received as well as made

Cellular phones have proven very popular despite considerable costs, generally several times more than for wireline phones. In most areas, cellular users pay not only for calls they make but also for calls they receive. The most spectacular growth has taken place in those few areas, such as Israel, which have encouraged cellular phones by setting costs only slightly higher than for wireline phone. In Israel, there is no charge for receiving cellular calls because the caller always pays, even if calling from a wireline phone.

Security

Telephone security has been a longtime problem. The problem is not limited to wiretapping or eavesdropping. Suspicion that operators were diverting calls to his competitors led Kansas City undertaker Almon Strowger to invent the first replacement for an operator—an electromechanical switch that could be controlled by a rotary dial telephone. Wiretapping has gone on for decades, often illegally. An entire industry has developed around ways to tap phones and countermeasures to detect tapped phones.

Cordless and cellular phones are even more easily tapped than wireline phones, and laws prohibiting such eavesdropping do not inhibit the practice, as many widely publicized instances have shown. Analog wireless telephones—both cordless and cellular models—are quite easy to tap; digital wireless phones are much more difficult to tap but by no means impossible. Accessories that scramble voice signals have been around for decades, but most work as analog devices. Digital scrambling is technically possible, but it would require complementary unscrambling by the cellular provider, because the cellular call is passed on as an analog signal to the telephone network, introducing a potential weak link. Some cellular phone companies may offer voice scrambling as a service.

Cellular Data Communications

Today there are several different methods of sending data over a cellular phone connection. The most familiar route to wireless data is over AMPS, the standard analog cellular phone system. You can use a conventional wireline modem; instead of plugging into a phone jack, you connect to a cellular phone with a data jack through an interface box or cable. These interfaces are designed to work only with specific brands of cellular phones; make sure to match an interface to your cellular phone before buying.

Interface boxes allow you to connect wireline modems to cellular phones

Two modem-cellular phone combinations have been designed to handle data. The AirCommunicator from Air Communications is a rather large cellular phone with a built-in modem. The size allows for modem circuitry and a

Using a Wireline Modem on a Cellular Connection

Most wireline modems have no special provisions for cellular telephone calls and usually work slowly or often not at all. You should connect while stationary and avoid peak call times—such as late afternoons—when the signal quality degrades. Motorola engineers suggest reconfiguring your modem for a cellular call:

- Make sure the software runs standard V.42 error correction.
- Set register S7—how long the modem will wait for a dialtone during call setup—to 90 seconds (the default is usually 50 seconds). Cellular calls can take a long time to connect.
- Set register S10—how long the modem will overlook a loss of the carrier signal—to five seconds or more (the default is usually 1.4 seconds). This can help keep the modem from dropping the call during a handoff from one cell site to another, which may occur even when you aren't moving.
- Lower the transmit level. Cellular phones try to make conversations as loud as possible to overcome noise, a strategy that tends to clip (distort) modem signals. Some modems support operation with an &L1 command that cuts the transmit level by 3 dB. But many modems lack any transmit-level control.

Even after adjustment, a cellular connection may not work well because the modem and cellular phone cannot work together optimally; a throughput of 200–300 characters per second (cps) is typical. (I use cps here rather than bits per second because the throughput is often much slower than the apparent connection speed.)

Modem-cellular phone combos control signal levels and monitor data exchange

battery for two-hour talk time. Motorola UDS offers external and PC Card modems specifically designed to work with its popular Flip Phones through its EC2 interface. The AirCommunicator and Motorola products are designed to increase speed and reduce the error rate by controlling the signal levels and closely monitoring the data exchange. Both products can work effectively, but their performance will necessarily vary with conditions. Both will operate as wireline modems as well.

A few modems go a step further by incorporating a special cellular error-correction protocol designed for radio communication. AT&T ETC (Enhanced Throughput Cellular) provides throughputs typically in the 700 to 1100 cps range, generally 20 to 30 percent faster than the older Microcom MNP-10 protocol. The rarely found ZyCellular protocol outperforms ETC in several tests. Naturally, all these protocols are incompatible with one another. Both sending and receiving modems must use the same protocol to gain real benefit (ETC installed in just the portable modem helps some); otherwise the modems will revert to V.42 operation or even basic V.32 without error correction. Very few receiving modems have any of these special protocols.

Some modems use cellular error-correction protocols, but most are proprietary and incompatible

There is a way around this problem. Some cellular services have installed modem pools in their cellular networks. When you dial data calls with a *DATA (*3282) prefix, the cellular system directs your call to a modem in its pool, which acts as an intermediary; it presents your portable modem with a cellular protocol and presents the wireline modem you are calling with a standard V.42 protocol. Primary Access Corporation, the main supplier of modem pool software, supports ETC and MNP-10. Some cellular providers charge for access to the modem pool (typically $5 a month). Even if you aren't using a cellular protocol, you may get faster throughput by using the modem pool, whose wired connections are designed for data rather than optimized for voice. The Primary Access modem pools also support TX-CEL, a modem enhancement process that works in conjunction with a cellular protocol or with plain V.42. Only a few modems have TX-CEL installed.

Modem pools can solve incompatibility problems

Most cellular providers will eventually install modem pools; for now you may need to switch cellular services in some areas. A few Public Data Network companies—which provide connection facilities for online services—have installed modems with a cellular protocol.

Compared to wireline, cellular data transmission is expensive

In the United States, cellular calls cost 25 to 75 cents a minute in addition to any long-distance and online charges, plus roaming fees if you're outside your normal service area. Thus, a megabyte costs $5 to $25 to transfer, compared to less than a dollar or even nothing for a local wireline data call. (Long-distance and online charges, if any, are not included in these cost estimates.) Some cellular providers offer discounted prices for data calls. Because you pay as long as you are connected, whether or not you are actually sending or receiving data, many users prepare their work in advance, connect for the minimum time possible, and get off.

Most cellular providers charge by the packet

AMPS is the only major wireless service that charges by the minute; other services break the data down into packets and charge by the packet. Since you are charged only for the packets of data you send and receive, there is no cost penalty for staying connected for hours at a time. Thus you can read and write your e-mail messages without needing to disconnect.

CDPD modems behave as if continuously connected but send and receive only when necessary

Another class of wireless service, **Cellular Digital Packet Data** (CDPD), operates in the cellular phone band but works very differently from a dial-up cellular call. Instead of maintaining a continuous carrier like a wireline or AMPS modem connection, a CDPD modem sends and receives only when necessary, piggybacking all the data on the existing cellular phone system by using the short quiet time at the beginning of ordinary cellular calls (but not the silent periods within a call) and hopping from one channel to another automatically.

When you establish a CDPD link, your modem registers itself with the nearest cellular transmitter. Every minute or two, the transmitter broadcasts a list of modems it has messages for; modems that have no pending messages can go to sleep until the next broadcast to save power. The peak nominal transmission speed is 19.2 Kbps; current tests suggest that the data throughput is about 800 cps in typical usage. With a CDPD link, your computer behaves as if it is continuously connected to a TCP/IP network via the Internet.

CDPD pricing encourages short messages. Many users may prefer CDPD for interactive messaging but go to an AMPS connection for long files.

Other Cellular Data Options

Other digital services are in the offing. The digital voice calls that the cellular providers are now promoting cannot be used for data, because the coding methods are designed exclusively for voice. In the future, a dial-up digital service for data over cellular frequencies could operate at fairly high speeds. Recently the FCC has reassigned and auctioned several other frequency bands that could provide data service.

Engineers are experimenting with new methods to achieve higher throughput rates

There has been talk about a method of sending data via CDMA that could reach throughputs of 64 Kbps. Although GSM and TDMA cannot yet compete with those throughputs, engineers are working on methods of transferring data over GSM and TDMA connections first at throughputs around 13 Kbps and then moving up to the 28 Kbps to 56 Kbps range by the end of 1998. Another jump for these technologies promises throughputs at 115 Kbps, achieved by a switch from circuit-switched networks to general radio packet service. This may happen some time in 1999, but it may require hardware upgrades as well as software changes. All of these technologies are in some sense stopgaps, with everyone looking forward to third-generation technologies that promise to provide throughputs up to 2 Mbps, although that level of bandwidth probably won't be available for several years at the earliest.

Pagers

Paging systems use multiple antennas and send short messages to battery-operated receivers

Pagers represent another point-to-point communications service, much simpler and cheaper than cellular telephones. A paging system uses several antennas to serve a metropolitan area; secondary antennas may be required in hilly or mountainous terrain. The antenna sends short messages to pagers, small battery-operated receivers with typically a one-line display showing from 10 to 20 digits. The simplest pagers, no longer popular, have no display at all; they simply sound a tone or vibrate on radio command. That's sufficient to tell the user to call a central office, for example.

Most pagers are receive-only; 10 percent support alphanumeric data

About 90 percent of pagers can receive only numbers. The rest can receive alphanumeric (both letters and numbers) messages up to 80 characters long. The vast majority of pagers are one way. Because they are unable to acknowledge receipt of a message, pagers typically send messages several times in case of reception problems.

Pagers work as a store-and-forward system: incoming messages are stored briefly and then transmitted. Pager messages are typically very short, most often only a telephone number.

Two-way pagers can respond to messages; responses are limited by keyboard availability

SkyTel operates a two-way pager network in the United States based on the Reflex technology from Motorola. The pager units can receive alphanumeric characters, but until recently have lacked space for a keyboard, so the reply is typically chosen from several possibilities contained in the message itself ("Can you meet at 9/10/11 AM?"). Two-way pagers with keyboards or those operating as accessories for laptop computers or personal digital assistants allow full replies. Only a two-way pager system can ensure that a pager message has been received.

Fax

Simple-to-use fax machines transmit data digitally

Facsimile systems send a copy of a paper document—a facsimile—over phone lines. Faxes have become an enormously popular use of telephone lines because faxing is so easy: you simply place paper in a fax machine, dial a number and press Start or Send, and that's all. Everyone knows how to dial a phone number, and most people are familiar with photocopiers; fax seems to work like a remote copying machine. Although fax is a digital process (in the sense that the fax machine scans the paper and coverts the image into a sequence of bits for the fax transmission), it usually seems more like analog because it is received in paper form. Unless you are using a computer fax program, you cannot refax the fax image without rescanning, which degrades the image quality.

Most computer modems sold today come with fax software, enabling them to send and receive documents in fax format. A fax modem is typically much fussier to use than a fax machine, although it does have its advantages. Because many documents originate on a computer, it's wasteful and time-consuming to print out a copy of a document, then fax it from a normal fax machine before recycling the original. With a fax modem, you can simply fax directly from the computer, eliminating one paper step in the process. In addition, receiving faxes can sometimes work better via a fax modem, because you can look at the fax before deciding whether or not to print it.[5]

Fax modems may be fussier but can save steps

Most fax machines and fax modems sold today use the V.17 standard for transmission. This provides for 14.4 Kbps of throughput, although fallbacks to lower throughputs, especially the 9600 Kbps that is common on the vast majority of machines, are available if the line quality degrades or the device encounters an older, slower model. As with normal modems, a pair of fax machines or fax modems transmit at the lowest common throughput. The Fax Standards table on the following page shows the throughput for each International Telecommunications Union (ITU) standard.

Most fax modems today operate at 14.4 Kbps

---

5. Sometimes printing is necessary for legibility, as the resolution on paper is higher than that on the screen.

Fax Standards

ITU Standard	*Throughput*
V.17	14.4 Kbps, with fallback to 12 Kbps
V.29	9600 bps, with fallback to 7200 bps
V.27ter	4800 bps, with fallback to 2400 bps
V.21	300 bps

In Conclusion

This chapter looked at various methods of point-to-point communications, primarily the public switched telephone network (PSTN) and the services that use it. Also examined were the technologies of cellular phone, pager, and fax communications.

A key concept is the distinction between circuit-switched networks, where you "own" an entire circuit for the duration of your use whether or not any information is flowing over that connection, and packet-switched networks like the Internet, where you're only "using" the connection when a packet of data is being transferred. Circuit-switched networks work well for voice conversations and other uses where the data transfer is constant. Packet-switched networks make far more efficient use of bandwidth in situations where data is bursty.

The Internet is essentially a network of packet-switched networks, and burstiness is just one of the many issues that complicate the way people think about bandwidth on it. The Internet with its bandwidth perplexities is the subject of the next chapter.

Chapter Eight

Bandwidth and the Internet

The most common measure of bandwidth over the Internet is bits per second, measuring how many ones or zeros can be delivered in a single tick of a clock's second hand. Multiplied by 1024 into kilobits per second, it's the measure of your modem's **throughput**. And as more and more people talk about bandwidth, articles in newspapers and general interest magazines toss out megabits-per-second figures with abandon. But rarely are the numbers as pure or simple as they may seem. Because the Internet's structure complicates bandwidth calculations, many claims amount to little more than stating that a car gets 100 miles to the gallon without bothering to mention that the measurement was taken for one instant while the car was running downhill.

Transmitting Confusion

As an example, let's look at so-called 10-megabit Ethernet, which is being used for some cable modem hookups to the Internet. It sounds as though it should deliver 10 megabits of data per second, which is a whole lot better than even the fastest standard modem. But that innocent 10-megabit number turns out to be anything but straightforward.

Housekeeping information attached to your data uses up a lot of those megabits

For openers, those 10 megabits include all sorts of mandatory housekeeping that the system must include along with real information, so that it can send the data along the right path. There are headers and footers that amount to substantial numbers of ones and zeros, not to mention some obligatory intervals of silence and additional delays that happen when one piece of data collides with another. The result is that though the raw bandwidth is 10 megabits per second, the throughput, or useful bandwidth—the amount of meaningful data you can transmit—is more like 4 megabits per second at best and rarely even gets to 3.

Multiple users on the system share the same bandwidth

This confusion of raw data with actual throughput is just the beginning. The second point is that all the users on the system share the bandwidth. If you're the first one to sign in and log on with your computer, all the bandwidth is assigned to you, and you've probably got yourself a great deal. But the picture changes radically once a hundred or a thousand of your neighbors sign up and start logging on when you do. All those data **packets** you want to send and receive have to share the bandwidth with the packets going to and from people down the block. Sooner or later, your connection may slow down significantly.[1]

Advertised speeds are seldom achieved

That is hardly the only misleading number in the world of digital bandwidth. As you've seen, so-called 56K modems sound as though they should work at 56 kilobits per second, but in the real world they cannot quite achieve that speed with information they're receiving, and when it comes to transmitting information, they work at roughly half that

---

1. Interestingly, although it's true that the effective bandwidth enjoyed by a single user drops as more users are added to the system, a shared portion of a higher-bandwidth connection generally averages out to be more valuable than a slower dedicated connection. In other words, it's often better to have 28.8 Kbps of a 1.54-Mbps T1 line than it is to have your own 28.8-Kbps modem connection. For a more detailed explanation of this, see Stuart Cheshire's article on bandwidth and latency in the online newsletter *TidBITS*. It's at *http://db.tidbits.com/getbits.acgi?tbart=00723*.

speed. And that assumes the best connections, "clean" phone lines that are often unavailable.

Let's try to clear up some of the myths.

Categorizing Internet Bandwidth

There are several different ways to categorize Internet bandwidth:

- **Bursty bandwidth:** Most of the information we deal with on our computers can be sent in short "bursts" with few problems. When you read an e-mail message, you don't really care whether it was transmitted in two parts or fifty or a thousand. Nor do you care how many packets came in at once. The important thing is that the message eventually arrived in your inbox and was assembled in the proper order. Within limits, the same is true of Web pages. It would be nice if a page just "popped" onto the screen in toto, but the fact that one part appears first, and then the next, and then the next is usually no big deal, particularly if the wait isn't long.[2] Even if you're downloading a video clip to be saved as a file for future use, bursty bandwidth serves just fine.
- **Streaming bandwidth.** Some forms of communication cry out for continuity. If the video or audio of a teleconference is interrupted, information is lost, and the conferees become annoyed. If you're playing an online game, you won't be happy if your responses arrive "sooner or later." In these cases you want bandwidth that's as close to an

2. Some early Web browsers collected all the data necessary to display a Web page before drawing it on the screen. Although this sometimes made the pages seem to display more quickly in their entirety, the technique proved to slow down most users, because people often prefer not to wait for all the parts of a page to display before clicking a link and moving on.

uninterrupted data stream as possible, which is why it's known as "streaming" bandwidth.

- **Guaranteed bandwidth:** If you're willing to pay for the privilege of making sure your packets are delivered cleanly and in timely fashion, it's possible to do so. A dedicated circuit should take care of the problem. For now, there's no way to guarantee bandwidth when your packets travel over the Internet, but that may soon change.
- **Broadcast bandwidth:** Another form of bandwidth that's just beginning to be possible on the Internet is broadcast. It's a one-to-many concept much like television broadcasting: data emanates from one point and travels to many other points at once. It's worth noting that unlike the other types of Internet bandwidth, this one offers no return **channel** for interaction with the sender. It's the couch potato of bandwidth.

How Much Does It Take?

The bandwidth of a POTS circuit is 64 Kbps

Let's take a quick look at the bandwidths in common use today for various types of activities and see how they might fit in on the Internet. A good place to begin is **POTS**—Plain Old Telephone Service. The reason it's a good place to start is that in the world of bandwidth, many things are expressed in multiples of a POTS circuit. As you saw in the "Telephone Bandwidth" section in Chapter 7, a single line amounts to 64 kilobits per second, or Kbps.

Slow Data Types

The following Slow Data Types table shows that a digital cell phone crams your conversation into just 8 Kbps, and a typical **PCS** (Personal Communication Services) cell phone will allow it just 9.6 Kbps. Remember, a POTS line gives your voice a munificent 64 Kbps, so if you've wondered why cell phones sound so much worse than standard phones, you've just learned a large part of the answer. You can do videoconferencing at slower data rates than 96 Kbps, but that rate has generally been considered the absolute minimum for anything that won't strain your eyes and ears.

Slow Data Types

Data Type	*Required Bandwidth*
Digital cell phone	8 Kbps stream
PCS cell phone	9.6 Kbps stream
RealAudio (poor sound)	14.4 Kbps stream
Minimal Web browsing	28.8 Kbps burst
Voice circuit (POTS)	64 Kbps stream
Minimal videoconferencing	96 Kbps stream

Slow Connections

The Slow Connections table shows some of the slower ways that people connect with the Internet. Combined with the previous table, it suggests that the Internet can serve adequately for some kinds of data. For example, if a digital cell phone can get by with less than 10 Kbps for voice, it seems reasonable that audio might work over the Internet and 14.4-Kbps modems—and, as RealAudio and other Internet systems for transmitting audio have shown, it can.

Slow Connections

Connection Type	*Bandwidth Provided*
Analog modem	33.6 Kbps
"56K" modem	45–56 Kbps
ISDN-1B (single circuit)	64 Kbps
ISDN-2B (dual circuit)	128 Kbps

Use of buffers gives the impression of streaming

But remember, a cell phone "owns" the circuit, so that information can be "streamed" without the worry about other data on the line. RealAudio and similar systems can't be sure that all the packets will arrive on time. So they use one of the many tricks that are employed to get around this problem: they make you wait a bit while they load up a bunch of packets in a **buffer** before the audio begins playing. The idea is to give the audio system a head start so that at any given moment there will be a bunch of packets ready to go, and while they're playing, enough more will come in to keep up the illusion of streaming. It works often enough, but sometimes the system will display a "net congestion" message and stop playing. Translation: the packets didn't come in fast enough to keep up; please wait until more arrive.

Even compression cannot compensate adequately for low bandwidth

Low bandwidth can be acceptable for Web browsing, because many pages are mostly text and can be delivered quickly. But large graphics files cry out for more bandwidth, and video absolutely hollers for it. Techniques for compressing pictures and video into smaller amounts of data can help, but video over slow connections where streaming is not guaranteed usually looks terrible no matter how much compression has been applied. In addition, the use of **lossy compression** (see the "Compression and Quality" section in Chapter 4) always reduces image quality somewhat.

Fast Data Types

You may be stunned by the amount of bandwidth required to deliver higher-quality streaming data (see the Fast Data Types table on the following page). With 384 Kbps, you get adequate video that fills only about a quarter of the screen and comes with decent stereo sound. For barely acceptable full-screen video with CD-quality audio, you need 1.2 Mbps. Even at 3 Mbps, the picture is no match for standard television. It takes about 6 Mbps to provide TV-quality video and 19 Mbps for high-definition television, which has been compressed before transmission. Normal broadcast television requires 23 Mbps, and uncompressed high-definition television needs a whopping 1.2 Gbps.

Fast Data Types

Data Type	*Required Bandwidth*
Quarter-screen video—good stereo sound	384 Kbps
CD audio—poor full-screen video	1.2 Mbps
NTSC—poor video	2–3 Mbps
NTSC—decent video	6 Mbps
High-definition TV—compressed	19 Mbps
Broadcast TV	23 Mbps
High-definition TV—uncompressed	1.2 Gbps

Fast Connections?

But where will you get all this bandwidth, and how much will you have to pay for it? How will we reach the higher bandwidths we all would dearly love?

The answer is that over the next few years, only a very few of us are going to have such access. Part of the problem is that the Internet today is nearing the end of a trial period where everyone involved made many blind choices and spent a lot of money without any real idea of what kind of return was available. I think that the problems of Internet bandwidth are going to continue for at least several years and perhaps much longer. And one of the biggest reasons is that bandwidth costs money. It may not always look that way, but it always does.

In the short run, few people will have truly fast access

How the Internet Works

Let's look at how the Internet works when you're at home in a suburb of Seattle and call up a page on the Web that happens to reside in a suburb of Boston (see Figure 8-1 on the following page).

Start with your computer and modem, dialing your ISP—your Internet service provider. We'll assume it's a 28.8-Kbps connection, which means it's an analog connection from your modem to the local phone company switch—where the

The "final mile" connection between your home and your ISP is analog

Figure 8-1 *A sample Internet session.*

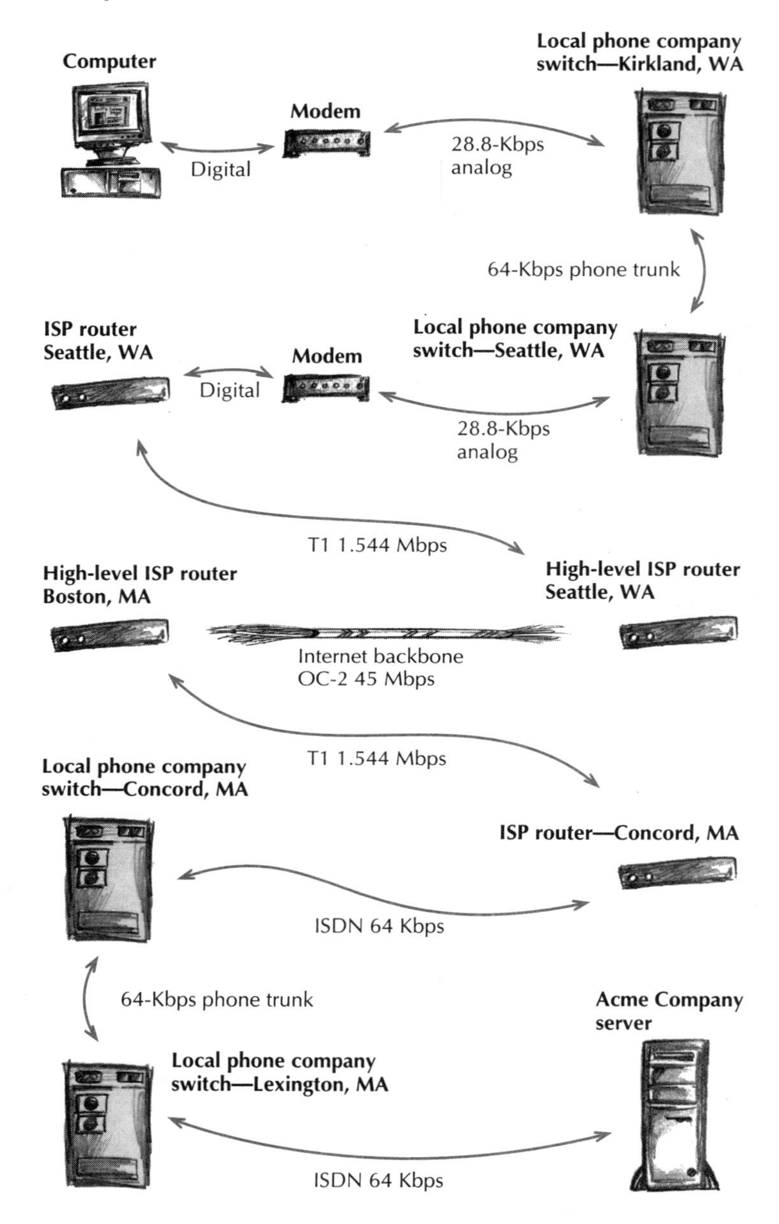

digital data that your modem turned into sounds is turned back into digital data in a different form. As you have seen, most calls today, whether voice or data, travel most of the way as ones and zeros; only the "final mile" from the phone company switch to your house or the business at the other end is truly analog. So even when you make a voice call, it typically travels most of the way in digital form.

ISPs pay by the foot for high-speed connections to the backbone

From your local phone company switch, the call goes to the ISP's local phone company switch, is converted to analog again, and is answered by the ISP's modem bank. In Seattle, as in many other cities, that modem bank is often located downtown. Why downtown? Because that's where the higher-level ISP that your ISP uses has fast digital connections to the Internet backbone. And because your ISP pays by the mile (actually, by the foot) to connect to its ISP, staying close makes sense.

Routers receive and transmit messages

Your ISP and its ISP typically use devices called **routers** to send your request out to the Internet backbone. Once your request reaches the Boston area, it goes off to another round of ISPs and eventually to a server that holds the Web page you requested. The information you want comes back to you, over the same route in reverse.

The fact that this works most of the time is quite amazing. But where can it go wrong? Where might there be bottlenecks? Just about everywhere.

Bottlenecks

Actual bandwidth is limited by the slowest connection

Let's start with the connection to the Internet service provider, generally a modem or Integrated Services Digital Network (**ISDN**) line. Although most people compare data connections on the basis of the raw data bandwidth, the raw data throughput is only one performance criterion. First of all, the commonly quoted data rate—14.4 Kbps, 28.8 Kbps—is simply a starting point. The actual data rate depends on several factors. As you saw earlier in this chapter, even if the modem or ISDN connection runs at its rated throughput, that's only the throughput for the final step, between the Internet service provider and the user. As you saw earlier, a typical Internet connection involves a dozen or more steps, and the actual bandwidth is limited by the slowest step, many of which are beyond the control of the Internet service provider or the user.

Let's look at some of these variables, paying attention primarily to modems, which are far more likely than digital connections to suffer performance degradation.

Modem Bottlenecks

Some modems cope with line noise better than others do

First, modems may not operate at their rated throughput. Current modems test the line for transmission quality during the initial handshake and during the connection. The throughput may be adjusted up or down accordingly. On many phone lines, the modems never actually achieve even 28.8 Kbps, typically getting somewhere between 21.6 and 26 Kbps. A few phone lines support 28.8 routinely; others may suffer along at 19.2 Kbps or slower. Modems vary in their ability to cope with noisy phone lines, and it is worth remembering that throughput depends on the modem at the other end of the connection as well. Although I've used 28.8 Kbps modems as an example here, the same issues apply to others.

Older serial ports on your computer may limit modem performance

Another bottleneck arises where the modem connects to the computer. The modem-to-computer interface may not allow full throughput. Serial ports in older IBM PCs and clones cannot run modems faster than 14.4 Kbps. Limitations in the PC Card interface of portable computers also often limit performance. The computer-modem interface bottlenecks depend on both hardware and software; one combination might work well, another might run 25 percent slower, even though both have the same nominal specifications.

It's worth testing to determine the best possible combination: in general, set your serial port at its highest possible setting, test, and work down until the connection works reliably. Although ISDN performance generally meets nominal specifications because the line is tested during installation, the ISDN interface to the computer may not run at full throughput if connected to serial ports on older computers, especially for 128 Kbps connections.

Virtually all modems built since 1990 have error correction and basic **lossless compression** built in. The error correction slows down the effective data rate by sending additional data that enables the modems on both ends to ensure that the communication has taken place correctly. Compression improves the data rate by recording the data in a more compact form; the original data is reconstructed at the receiving end.

Error correction reduces the data rate, but lossless compression usually makes up for this

The amount of compression possible depends on many factors. In brief, text is typically reduced to half the original size, while graphics vary tremendously. Data that has already been compressed is not usually compressed much further. The error correction and compression require that the modem at each end of the connection be equipped with compatible features, which are universal among 14.4 Kbps and faster modems intended for POTS circuits. A few modems offer proprietary compression variants that are supposed to improve data throughput, but only if the same type of modem is in use at both ends of a data connection.

Latency

There's another subtle problem with modems that can reduce performance significantly, especially in comparison to digital connections. Although most people think of downloading a large file as a single event, many very small control messages actually go back and forth during that large file transfer. In short, performance depends not just on the raw data rate but also on how quickly a single tiny packet can be transferred. The minimum time to transfer a packet is

Transmission of small control messages has an impact on performance called the **latency** of a device. Unfortunately, once a delay has been added by latency, there's no way to remove it.

Millisecond delays add up quickly to become noticeable

It turns out that modems have horrible latencies, somewhere around 100 milliseconds, 300 times worse than Ethernet latencies. Modems were designed for streams of data going between a terminal and a mainframe, and thus they have built-in delays before they can start compressing and sending a block of data. Those delays, originally designed to identify when someone stopped typing, are no longer necessary, but they add overhead to every piece of data a modem transmits or receives. The connections to the computer generally add delays as well, and although all of these delays are measured in tens or hundreds of milliseconds, they add up when applied to every piece of data a modem sends.

Set your serial port high or switch to digital

Not only is there no way to eliminate latency once it has been added, but, short of complaining to the vendors to get them to design modems with lower latencies, there's little individuals can do to reduce the latency of their modems. In fact, just about the only thing an individual can do is make sure the serial port is at its highest reliable setting; that reduces the computer-to-modem delays slightly. Switching to a digital connection also solves the problem, of course.

Latency becomes most important with interactive communication

You may wonder why latency would matter much. In reality, the larger the piece of data transferred, the less latency does matter, because the raw data rate becomes much more meaningful. However, in any situation that requires interactivity, such as voice or video communication or an interactive game, latency becomes tremendously important because most of the data is comprised of tiny messages back and forth, updating each person on the status of the call, the videoconference, or the game. A certain amount of latency is unavoidable due to constants like the speed of light in fiber-optic cable or the speed of electrons in copper wire, but modem latency is about a thousand times worse than the latency introduced by the speed of light.

Conversions

Because modem calls are analog, bandwidth suffers if the line is noisy, as it might be if you have old wiring or if squirrels have been gnawing on the cable. All those conversions between analog and digital and back again covered in the "Public Switched Telephone Network" section in Chapter 7 turn out to reduce bandwidth; so-called 56K modems achieve their extra speed by eliminating the conversion on the ISP's end. But what if your modem can't connect with your ISP in the first place, thanks to a busy signal?

Noisy lines and data conversions eat up bandwidth

Phone Company Limitations

Let's look at it another way: What did you think you bought when you ordered your "unlimited" local calling service from your local phone company? What did you think when you bought "unlimited" Internet service from your local provider for $20 a month? It turns out neither one is truly unlimited, and neither one truly makes sense.

Phone company switches have limited capacity

Assume for the moment that your ISP has plenty of modems to take your call. That doesn't do you much good if they're all busy. There's a similar problem down at the phone company; a typical phone company switch can handle only about 10 percent of the lines connected to it at any one time, though the percentage is higher in some business areas with heavy phone traffic.

Most of the time, 10 percent usage provides a nice cushion, because the phone system is designed and built for voice calls. But when there's a public emergency like an earthquake or a "call home" holiday like Mother's Day and everybody picks up the phone, the system simply can't handle the load. Depending on which end has the problem, callers will either hear no dial tone or calls that end in "all circuits busy" messages.

Phone company capacity may not be able to accommodate Internet usage

The Internet is beginning to cause usage patterns that look like public emergencies. The average voice call in California lasts 4 minutes; the average voice line is in use 22 minutes a day. But for lines connected to the Internet, the average time for a single call is 22 minutes, and the average line is in use 62 minutes a day. This heavy usage can eat up the 10 percent of lines that are available at any one time.

If you've got "unlimited" service, why worry? Some kids find that their ISP's lines are busy after school, so they dial in before they leave for school in the morning and stay connected all day so they can check their e-mail messages when they come home. Taken to extremes, the result could be a neighborhood without dial tone.

"All circuits busy" is not unusual at peak times

At the other end of the line, where the ISP's modems live, the available lines may run out too, particularly if a lot of ISPs share the same switch. In Seattle, that's exactly what happened in January 1997. At peak times, 88 percent of calls to many downtown exchanges encountered "all circuits busy" signals. To the standard peak times around 10 AM and 2:30 PM, a new one was added: the late evening, from 9 to 11 PM. What's more, many Internet calls were placed by computers set to redial until connected, further jamming up the system.

Traditional planning techniques have become obsolete

Managing capacity is one of the phone companies' most important jobs. Anyone can overbuild a system, but that would add costs that are unnecessary for handling the vast majority of situations. From an economic point of view, if the phone system never jams up, it has too much capacity. An ideal system jams up only on very rare occasions. But the Internet has made traditional telephone planning obsolete.

Downsized phone systems deserve some blame; fax machines and cell phones do not

Critics point out that many factors, not just the Internet, conspired to produce the Internet jams. Phone companies downsized in the early 1990s, betting on slow growth in the coming years. They persuaded utility commissions to change from "cost-plus" pricing to a new model in which they could pocket the savings from greater efficiencies. Pacific Bell increased its profits by operating its system at 95 percent of capacity rather than the 85 percent it had used before. Some Internet users are in denial: they insist it's the fax machines and cell phones that are causing the problems, not their Internet connections. But it's clearly the Internet that's the worry; fax machines and cell phones use the phone system in patterns that resemble typical voice calls.

How Will Capacity Increase?

It is becoming clear that telephone capacity will have to be increased. But that costs money. Who will pay? The phone companies don't want to pay, although all have made emergency investments in the past couple of years. Their argument is that they are regulated on the basis of voice calls. With unlimited service, they get no extra money for all the Internet calls clogging their facilities.

Telephones are a gateway to the Internet

Internet calls essentially use the local phone system as a gateway to another service—the Internet. In that, they are similar to long-distance calls, where the local phone company provides the gateway to the long-distance carrier. The long-distance carrier pays the local phone company for the use of the local phone system, and pays by the minute. The amount of such payments is the subject of an ongoing dispute between local and long-distance phone companies. A portion of the reimbursements goes into a universal service fund, so that rural areas can get low-cost phone service.

Modem calls were not required to pay carriage fees

But in 1983, the FCC exempted modem calls from paying the local phone companies for carriage. The idea was that modem calls were a new service and needed some encouragement. In 1997 the FCC decided not to impose such fees, but the issue is unlikely to go away. Expect much more argument on this topic. If fees are imposed, expect a row over fairness issues such as whether people in rural areas should get a subsidy for Web surfing the way they do for their phones.

ISPs have a good deal

The ISPs don't want to pay, because they get an unusually good deal. The business phone lines they buy from the phone company are essentially flat-rate lines even where no flat-rate service is available for business phones. The reason is that the ISPs' modem pools make no outgoing calls. For a single basic service fee, they can receive all the calls they can handle, even if the line is in use 24 hours a day. Is that fair?

Users don't want to pay more, either

And users? We don't want to pay, either. We like our "unlimited" local phone service and "unlimited" Internet service, even if in the long run we waste valuable time because we and millions like us clog up the phone lines and the Internet because there is simply no incentive not to do so.

ADSL may prove to be the solution

Asymmetric Digital Subscriber Line (ADSL), discussed in the "Digital Subscriber Line Service" section in Chapter 7, is in some ways an attempt to resolve this problem. It manages this by splitting the voice and data portions at the switch. Voice calls travel out onto the phone companies' voice networks, whereas data is routed into the phone companies' high-speed data networks. It's a much more efficient use of existing facilities, and the cost savings may be so great that phone companies will move more quickly than they traditionally do when introducing new technologies.

ISP Traffic Jams

Failure-to-connect rates can be as high as 50 percent for major ISPs

An ISP has only a limited number of phone lines, typically about five for every hundred subscribers. When America Online got into big trouble in December 1996 after switching to unlimited service, it had only 3 percent coverage and quickly began building out to 5 percent. ISPs rarely publicize their modem coverage percentage, in part because even 5 percent means that at peak hours, some people will not connect, even if there are no problems with the phone system. Surveys show that the combined failures to connect—busy signals, circuits busy, failure of modem connection, and so on—run between 20 and 50 percent for the major ISPs.

But suppose you do manage to connect to your ISP. Are you getting the bandwidth you paid for? Can you really expect even the 28.8 Kbps your modem can deliver? Probably not. One reason is that, unbeknownst to you, you're sharing bandwidth with the ISP's other customers. The connection to the ISP's "front door" is yours and yours alone, but the connection out the "back door" to the Internet is divvied up among you and every other customer who happens to be online at the same time.

Your ISP's connection is shared by many

Suppose the ISP buys a high-grade T1 connection to its higher-level ISP. That line offers 1.54 Mbps, the equivalent of 24 64-Kbps voice circuits or roughly 50 28.8-Kbps modem lines. But how many modem lines does the ISP allow to connect to the single T1 connection at the same time? Many more than 50.

This actually makes sense. On the Web, much of your time may be spent reading rather than actively sending or receiving bits of data, and the same goes for the ISP's other customers. But what is the reasonable oversell factor? What's a sensible ratio of bandwidth the ISP has sold to bandwidth the ISP has bought? Whatever the answer, it's a number that ISPs don't want to talk about. In researching this book, I could not find a single ISP willing to divulge a number for the record.

ISPs have to oversell their bandwidth

For their most important business customers, the best ISPs apparently use an oversell factor of five. In other words, they might sell the equivalent of five T1 connections and aggregate that data into a single T1 line that connects to the Internet. For modem customers, the oversell factor is clearly much higher, probably somewhere between 10 and 30. The result is that users may find themselves with slow connections, particularly when some users hog the system with streaming data or big downloads. The upshot is that with most ISPs, even 28.8-Kbps performance over a modem line isn't guaranteed.

High oversell rates lead to slow connections

ISPs provide the minimum acceptable service

Few providers make much money on the typical $20 account. Some have suggested that they break even only as long as the customer logs on for no more than eight hours a month, particularly because dial-up users tend to be expensive to support. To be cynical, there is only one fundamental principle when it comes to selling bandwidth: deliver the minimum that the customer will accept without canceling the service. Because more bandwidth is a cost that comes out of the bottom line, no other principle makes any financial sense.

For users, there is an equally cynical principle that is a bit like dividing up the water in a river. As long as pricing is not related to usage, everybody has an incentive to grab as much as possible. It is a crazy situation, because the average user subsidizes the heavy user. It is an electronic version of what microbiologist Garrett Hardin has called the "tragedy of the commons,"[3] where unlimited grazing permitted on a public common inevitably causes overall damage to the community because there is no incentive for individuals to moderate their behavior.

The situation becomes truly cynical when you consider the idea of independent surveys of ISP performance. If people suddenly flock to the "best-performing" ISP, it will quickly be overwhelmed, at least in the short run. If you find a good ISP, it is in your interest not to tell anyone. In the short run, the more people who compete for the ISP's limited resources, the less there will be for you.

Other Bottlenecks

Problems at local phone companies and ISPs are by no means the only culprits contributing to slowdowns on the Internet. Because the Internet is a packet-switched network, it's share and share alike, and you are sharing with millions

---

3. "Tragedy of the Commons" in *Science,* 11 November 1968, 1243–48.

of other users at once. At peak hours, when everyone is contending for the same limited resources, things can slow down almost anywhere. Here are a few of the problems:

- **Servers.** The computer that holds the data you want is known as a server, and the speed at which it can deliver data is constrained by many things, virtually none of which you control. Its hard drive, processor, input/output system, and connection to the Internet all affect how fast it can send data to you. Many Web sites share their servers with other sites; if a popular site resides on the same machine as the little site you're looking for, that too can slow down your access.
- **Internet infrastructure.** The primary interconnections that carry Internet traffic are known as its "backbone." Again, resources here are far from unlimited, and the packets you send and receive must compete with everyone else's. At peak times, delays in the backbone and other Internet plumbing such as routers and name servers can hamstring everyone's bandwidth.
- **Protocols.** The peculiar method that the Web uses currently requires each separate element of a page—graphic, picture, map, and so on—to be delivered in a separate transaction that adds overhead beyond delivering the actual data. The result is that separate elements can reduce effective throughput.[4]

---

4. Although there is additional overhead associated with separate transactions, which reduces the effective bandwidth, the use of separate transactions for different parts of a Web page, say, can increase the perceived bandwidth because the user receives at least some useful data more quickly than if he or she has to wait for the entire page to load.

Alternatives to Voice Lines for Reaching the Internet

If you work in an office, you may well have a T1 or T3 Internet connection that you share with your colleagues. If enough people use it at once to stream data, even that kind of connection can become sluggish. Until fairly recently, the "final mile" option for most people in their home or office was limited to a phone line and a modem, which was virtually guaranteed to work at a snail's pace. Now that has begun to change. Several of the following alternatives are available today, and the others are likely to arrive in the very near future.

Integrated Services Digital Network

ISDN offers only modest advantages over POTS

As you may remember from the "ISDN" section in Chapter 7, ISDN was originally proposed in the 1960s as a circuit-switched voice system. In the United States today it is typically used for data. Its 2B+D system includes two so-called B channels, each of which amounts to a single 64-Kbps voice line; the D channel sets up calls quickly and can manage a modest data stream. Because ISDN was designed for voice and is circuit-switched, it offers only modest advantages over standard POTS (Plain Old Telephone Service).

You can bond two channels together

In many cases you can "bond" the two B channels together to achieve 128 Kbps over what looks to the computer like a single connection. And there's another advantage: although standard modems take a long time to set up a call, ISDN makes connections in a matter of a few seconds. However, in the age of so-called 56K modems, ISDN isn't very exciting, particularly considering the premium price that phone companies and ISPs often exact for it.

Circuit-switched connections are not automatically continuous

Circuit-switched connections like ISDN have an additional disadvantage: unless you are willing to leave your phone on all the time and hog a circuit, they are not "always on." The other ways to the Internet are packet-switched all the way into your home or business. Because they offer a continuous presence, you never have to dial in to see if you have an e-mail message.

Cable Access

Cable modems have been highly publicized but slow to roll out to the real world. One design, which minimizes upgrading of cable TV systems, uses the cable to deliver inbound data at a high speed and the phone system to send data back at a crawl. Of course, that's not nearly as attractive as such systems as @home, which offer a full two-way service. Typically, a 10-Mbps Ethernet network is at the core of such a scheme, meaning that the actual throughput is no more than 4 Mbps, and all of it is shared.

New systems offer two-way service

One of the tricks @home uses to improve performance is something called a proxy server. Essentially, you can think of it as a big hard disk drive that stores the most popular Web pages that people have requested. The first time someone requests a particular page, the proxy server offers no performance advantage. But, the next time someone asks for that page, the server can deliver it without first having to retrieve it from the Internet.

@home uses a proxy server to minimize the amount of back-door bandwidth needed

Not only does this improve performance, but it also minimizes the amount of back-door bandwidth to the Internet that the provider needs; in fact, America Online and other large ISPs use a similar scheme. At the time of writing, the proxy server in a West Seattle trial of @home held 9 GB of information, and, surprisingly, held about 50 percent of the pages customers requested. What was on those pages? Mostly pornography.

For the moment, a typical session with a cable modem will be significantly faster than a dial-up session over a POTS line. How fast it will be when thousands of households hook up to the network remains unclear. Thus far, most

Future cable modem service will likely integrate the Web and digitally delivered television

publicized projections of cable modem service acceptance have turned out to be wildly optimistic. The upgrades that many cable systems will require are expensive, and the financial returns remain unclear. However, as cable systems increasingly change their systems from analog to digital, providing Web access will become more commonplace, and much of that access may take place over set-top boxes that integrate Web facilities with digitally delivered TV. Unfortunately, cable Internet access will likely come with the fabled customer service of the cable companies, which only rarely manage to deliver decent TV pictures.

Digital Subscriber Line Services

xDSL schemes vary

Like ISDN, the Digital Subscriber Line (**xDSL**) systems discussed in the "Digital Subscriber Line Service" section in Chapter 7 use the standard copper wire available for most telephones. But the "x" signals one of the problems: there are several different and incompatible schemes for delivering high-speed access, and the speed itself is an issue. It is available today from scattered phone companies at wildly differing prices, which typically begin at $50 per month.

The xDSL systems use standard components but many different flavors. Here again, the proxy server and ISP "back door" bandwidth will be crucial. No one is likely to offer multiple megabits per second out the back door at low cost, unless the costs of high-speed T1 and T3 lines magically decline.

T1 Apartments

Sharing service can be cost effective

A few new apartment buildings have built standard T1 lines into their infrastructure. Everybody in the building is networked together and shares the line. Including ISP costs, the service can end up costing around $200 to $250 per month. Again, this is a shared service: if someone in another apartment is using streaming video or sets up a server for a busy site, your share of the pie may dwindle beyond recognition.

Wireless and Cellular

A company called Metricom has deployed a wireless service called Ricochet that operates in the unregulated 900 MHz band and delivers throughput comparable to 28.8-Kbps modems. The service is currently available regionally in Washington, D.C., the San Francisco Bay area, Seattle, and some airports for about $30 a month plus the special wireless modem. The company has said that it will have a service that works at ISDN rates of about 128 Kbps in 1999. Despite a recent infusion of cash, it remains unclear whether the company has the wherewithal to create a truly nationwide network, and if so, how soon it might be available.

Some services are available locally

Sending data over wireless cellular and PCS systems tends to be both expensive and slow, not to mention unusable in a moving vehicle. That said, it may still be useful for people on the go who need to keep in touch with their e-mail.

Satellite Services

Today, you can access material from the Internet with the help of a special dish antenna that may or may not also capture DBS satellite TV signals. This scheme has several drawbacks: you share the satellite with every other user over a huge area, the path to upload data is through a standard telephone, and the service is quite expensive.

Internet satellite connections are inbound only, and expensive

Two new systems that are expected to be deployed in the near future are called Teledesic and Iridium. Although they are likely to be popular with businesses in remote areas of the world, they should have little impact on all but the most affluent urban Internet users. Outside the United States, they will typically be configured like wired phone systems and operated by local telephone monopolies, which are unlikely to permit a competitor to enter the market and undercut their prices.

Spread-Spectrum Wireless

Throughput may fall with too many users

Granularity is the ability to deliver distinct digital data to finer and finer physical locations—ultimately an individual office or desk. This is hard to do, but technically possible in principle. A frequency-hopping wireless system has been advocated by technology author George Gilder in *Forbes ASAP* magazine. How well it might work remains to be seen.

A very dense radio system is unlikely to be efficient for offices. At some point the noise level may rise and throughput may fall in an ungraceful way. With any cellular scheme, there is usually some maximum number of people a particular cell can serve over a particular area. Like many other systems, this one is likely to work much better as long as not too many people use it at once. But then the costs of deploying such a system rise significantly.

Theory vs. Practice

One problem with many of the systems that have been announced is that they have not yet been deployed or even tested under full loads. How they perform in the real world may be very different from how they perform in theory. How they will perform when lots of people want to use them to access the Internet at once may be very different from the early days, when only a very few pioneer subscribers are interested.

What Will It Cost?

T1 speeds are most likely achieved only on portions of the transmission path

What might it cost to get faster Internet service? At the moment, true T1 service costs $1,100 to $2,500 per month plus the phone line charges, although there is a tremendous variation in pricing. So how can anyone pretend to offer T1 speed for less than $100 a month, as some xDSL vendors are promising? One cynical answer is simple: deliver service that only rarely lives up to what it promises. Unless the basic price of true high-speed services drops radically, which in a market that mixes regulated services with deregulated and unregulated ones remains problematic, promised T1 speeds are likely to apply only to some small portion of the data path.

There's another question with an unclear answer: How many people are willing to pay $100 a month for 1.5-Mbps service? Or even $50? It seems quite clear that most people aren't willing to pay $100 per month for any level of Internet service, so unless the overall cost can be in the $50 per month range, most consumers probably aren't interested. This uncertainty is likely to prevent fast adoption of any of the high-speed services.

Users are not likely to pay more than $50 per month

Capitalism is not terribly interested in philanthropy. At one phone company the mission statement was simple: "maximize shareholder value." There was not even lip service paid to the idea of providing communications. Until companies have a better idea of how many people are truly willing to pay for fast connections, the investment to make such connections possible will be slow.

Companies don't care about the users anyway

Other Cost Factors

There are several other factors that may have huge impact on Internet bandwidth costs and service. The first is the availability of "unlimited service" for $20 a month. Unlimited doesn't mean much if you can't log on, or if when you do get on it's creaky; but it may be acceptable anyway if you're not too fussy. The mere availability of this cheap alternative is one less incentive to provide good service. Many people may find that for what they do—e-mail and occasional browsing—it is not worth paying a nickel more, let alone $250 or more a year.

"Unlimited" means nothing if you can't log on

So-called RSVP service for guaranteed bandwidth that you can count on may be another factor. The costs are not yet known, and some observers propose auctions, but RSVP

The price of guaranteed service for special data packets is unknown

packets will in effect get to elbow others out of the way in much the way first-class airline passengers board first. If RSVP proves popular, the quality of service for others on the same wire may decline. RSVP may become a higher class of service, separate from the bandwidth. If enough people pay for RSVP, conceivably, competing services will be squeezed out.

Charges based on usage might be fairer

One possibility is that we'll end up with a tiered service model, where, in essence, you pay for what you use. Although people generally hate the concept of such models, they actually make sense. If all you do is check your e-mail every day or so, you'd probably end up with a tiny bill, whereas the person who spends hours each day playing online games would bear the equivalent costs.

ISPs used to pass each others' packets along without charge

A final variable in this equation is a tricky issue known as **peering**. Essentially, all ISPs must connect to the backbone, which is run by a relatively small number of very large ISPs, also often called network service providers. Originally, peering referred to the process whereby everyone agreed to pass everyone else's packets along for free, which made sense as the Internet slowly moved from being an academic resource to a commercial venue.

Now some ISPs charge one another for packet traffic; transmissions from small ISPs suffer

As the amount of traffic has increased, however, some of the large network service providers have changed their policies, accepting packets only from other large network service providers and smaller ISPs who are their customers. Small ISPs who can't afford to buy service directly from one of these larger companies could find their packets relegated to much slower routes, and in the worst situation, find that no one will accept their traffic on the backbone at all. Needless to say, such a situation would be untenable, and the small ISP would have to figure out some way of playing up to the network service provider. This is yet another factor in the coming consolidation of the ISP market, as small ISPs find that they can't make money charging $20 per month to their customers and paying high prices to the network service providers for access to the backbone. In the end, this will probably drive bandwidth prices up more.

But on the other side of the coin, huge volumes of bandwidth are coming online every month. Companies like Qwest and Level 3 are laying thousands of miles of fiber-optic cable with immense carrying capacity. If the business to fill these capacities does not materialize quickly, bargains may abound, at least at the wholesale level.

Broadcasting Data

European teletext uses the VBI portion of the television signal

Systems that borrow some of the techniques of the Internet are poised to deliver bandwidth in other forms, most of them one-way, as the collective description "data broadcasting" would indicate. Their granddaddy, popular and well-established in Europe, is the system called teletext, which broadcasts text pages over the **vertical blanking interval** (or VBI), a portion of the television signal that is also used for closed-captioning for the deaf. The VBI is broadcast while the electron beam jumps from the bottom of the screen to the top; you can see it as the black bar that rolls up when a TV set's vertical hold is out of whack. It can deliver a surprising amount of data. (For more details, see the "Datacasting in the Television Channel" section in Chapter 6.)

New systems, using VBI with computers and software, can stream data at 28.8 Kbps

Teletext is fairly primitive, delivering mostly blocky characters designed to be read on a television screen. The newer systems that are about to appear will be more sophisticated. One, called Wavetop, uses the VBI to deliver information such as Web pages. It requires a video tuner in a computer and uses software to decode the signal. The system can deliver a continuous stream of data at about 28.8 Kbps, which is easily enough to offer sidelights of information such as sports or stock statistics that can appear on the screen in sync with the program you are watching—presumably with their own ads.

Downside: cable services can strip out VBI

There are problems with delivering data via the VBI. One is that cable services are permitted to strip most of that signal out and deliver whatever they want instead. That's why a variant of the scheme would use an entire TV channel to deliver nothing but data at far faster speeds. That is theoretically safer than VBI, at least until cable systems begin compressing channels digitally to cram more of them into existing capacity.

Upside: VBI provides opportunity for storing large amounts of data for fast retrieval later

But because the VBI can deliver about 10 megabytes per hour, or 240 megabytes per day, much of the data stream might have no relationship to the programs you are watching, being stored instead on your hard disk drive for quick retrieval later. The same goes for a TV "data channel." Yet another scheme involves a satellite dish connected to a computer or set-top box with a big hard disk, which is presumably even more reliable, as the sender of the information should have more control over it. A single satellite transponder can send many gigabytes per day.

Use of persistent storage gives the illusion of bandwidth

Suddenly these become potentially interesting examples of using "persistent storage"—the hard disk drive—to create what might be called "pseudo-bandwidth." Information you have specifically requested and news that the system thinks you might want to have on hand are delivered automatically and stored on the hard disk drive. When you request these bits of information—for example, when you wake up in the morning—they are right there on your machine, so access is almost instant, as though the bandwidth were virtually unlimited. With this scheme, high-quality audio and small amounts of video become feasible. One popular service might be the delivery of software upgrades and bug fixes.

Persistent storage methods insure that data is not entirely fresh

The problems include guaranteed delays; your online newspaper, delivered in the wee hours of the morning, will not be quite as fresh as the morning TV competition. If you want information that does not happen to be on the hard disk, you'll have to access it separately via more conventional, and probably slower, means. Because data broadcast is one-way only, it does not include specific items meant for you alone, like e-mail. And what happens if your hard disk crashes? Who will you trust to take over part of your computer? And what if one of those upgrades is actually a virus?

So far, Internet "push" media providing information to users without their specifically requesting it have not been spectacularly successful. The big question for data broadcasting is whether it can find information that is compelling enough to make people want to use it. It would seem to make a lot of sense for general information like news headlines, weather, school closings, and traffic reports—precisely the sort of stuff that teletext has delivered for years. Beyond that, it is not clear how well it will work, and as with other broadcast media, striking a balance between local items and items of a broader interest may be tricky.

What Will Fill Our Bandwidth?

Content will come at a price

Bandwidth, as I have said before, does not measure the value of the content it delivers. But the only thing that makes bandwidth valuable is that content. On the Internet, content has traditionally been free, but I believe we are almost certainly going to have to start paying for it.

Not much Internet content is suitable for children's schools

When I served on the Committee for the Public Understanding of Science for the American Association for the Advancement of Science, I systematically looked for science information publicly available on the Web, especially material that was suitable for schools. And after a lot of looking, I came to question why people like Al Gore call for connecting every school to the Internet.

Aside from some nice pages from NASA and the Hubble telescope and a few animations, there is virtually no science content on the Internet that is appropriate for kids. For most schools, three or four shelves of really good books would be a much better bargain than Internet access. Since the Internet is good for e-mail, many have promoted collaborative projects with classrooms halfway across the country. But if collaboration is important, why not start with the classroom next door?

Good content, like real news, is hard to find

Without some form of reimbursement, there is little incentive to do good work. I propose a standard that will tell us when the Internet has arrived as a truly valuable information source: when we pay as much for Internet-delivered content as we do for our connection itself. Today, cheap, low-quality "information"—including everything from the self-promotion of self-styled healers to corporate press releases masquerading as news—is available all over the Internet. Real news, delivered by real journalists, is harder to find.

Newspapers, hoping to attract advertisers, are offering free content online

For the moment, many newspapers are offering versions of their products free of charge on the Web, because once they have gathered the information, the incremental costs of putting it on the Web are relatively small. So far, only the Wall Street Journal among the major newspapers has been willing to charge readers to view its online information, and even then it charges less than it does for a subscription to the daily paper. The hope is that the Internet version of a newspaper or magazine will not cannibalize paying subscribers or that it will attract enough advertising to make up the difference, but so far that is only a hope.

Delivery of the information we get today in books, unencumbered by advertising, remains virtually nonexistent online, in part because of very real problems of piracy. It is possible that a Gresham's law of information may apply: cheap, low-quality pseudo-information may drive out the good information that is expensive to collect, edit, and verify.

Many people believe the only viable system for charging for information on the Internet is a micropayment system, where you would pay a tiny amount, perhaps one cent, per Web page. The idea is that the amount must be small enough that you wouldn't even think twice before paying (or rather, before letting an automated system pay for you). It has been pointed out that no one thinks twice about turning on a light on entering a room, and many people don't even think about turning the light off when they leave. Electricity costs money, but so little for each individual action that it's not worth thinking about. In contrast, when something costs 10 cents per minute, like a long-distance phone call, people start to think about how long they're on the phone, and international phone calls, which can be 40 cents per minute or more, tend to be closely monitored.

A micropayment system might be palatable

Unfortunately, a micropayment system requires a complete infrastructure for handling credits and debits, and that infrastructure must be incredibly efficient so as not to eat up each charge in processing fees, as would happen currently with today's credit cards. Micropayments are still several years in the future, but given that Web subscriptions haven't proven particularly successful, and because advertising has turned out to be viable for only the largest of Web sites, micropayments may be the only long-term solution.[5]

Micropayment systems will have to be efficient

5. Jakob Nielsen, a Sun Microsystems Distinguished Engineer, has written about micropayments in his Alertbox column *The Case for Micropayments* at *http://www.useit.com/alertbox/980125.html*.

Afterword

The Future of Bandwidth

The advantages are too great in virtually every way for the storage and transmission of information not to become almost exclusively a matter of ones and zeroes. Digital technology will ultimately prevail, and bandwidth will be measured accordingly.

But it is worth remembering that although the world of information is becoming increasingly digital, it is still primarily analog in the way it interfaces with us. Not one of our senses operates digitally, which means that the capture and expression of information will at some ultimate level remain analog, be it the sound waves of a song or the light waves of a picture. The implant offering a direct link between our brain and nervous system and digital data has been a staple of science fiction for many years, but it is safe to assume that it will remain in the fictional domain for years to come.

That striking potential use of digital bandwidth may not come to pass in our lifetimes, but many things almost as startling are likely to appear sooner. Telepresence will allow us to see and interact with people and things thousands of miles away, as though they were in the room with us. Virtual reality will let us experience real places we do not have the chance to visit, plus synthetic places that exist only in cyberspace.

There will be limitations. One, as we have discussed, is latency. The merest delays in transmitting and processing data are annoying enough to spoil the illusion of interaction. Latency can spoil the effect of a complicated virtual-reality simulation just as surely as it can negate the possibility of long-distance surgery.

How much bandwidth will there be as the twenty-first century progresses? There is no question that for some people, in some places, bandwidth will be abundant and possibly cheap. As is the case today, you will obtain as much bandwidth as you need, provided you are willing to pay for it. That points to high-tech meccas like Silicon Valley, Seattle, and Boston as early candidates for high bandwidth widely available to the public at large.

Bandwidth does become cheaper when the cost is spread out. If someone runs a high-speed cable down your street and everybody takes advantage of it, the cost of amortizing that cable declines dramatically. If you are the sole user of that cable, the costs will likely be higher. In areas of high use, special forms of shared bandwidth may arise. In many less fortunate areas, digital bandwidth piggybacked onto upgraded TV and telephone cables that others are using for their original purposes may well remain the wave of the future.

It would be a great mistake to underestimate the importance of economic factors in the bandwidth we will ultimately enjoy. The digital haves are unlikely to subsidize the digital have-nots to the same extent that we have seen in the world of telephony, where tariffs were developed to ensure that even people in rural communities were served at affordable rates. And there is likely to be great turmoil over bandwidth turf: every player is likely to try to control areas where control is possible and economic benefits can be derived. Everything from software and hardware standards to the nature of secure business transactions will be up for grabs in this digital future.

Some things will remain analog. Will there ever be a time when all human knowledge and content will be available in digital form? The answer is clearly no, because converting the vast library of books, newspapers, records, magazines, films, and tapes is too expensive for the limited demand that exists. But there will be enough content in digital form to supply what people want, as witness the conversion of many (but not all) recordings from analog records and tapes to compact discs. The unfortunate aspect is that those print materials that aren't converted into digital form will largely slip from all human access except for a band of intrepid specialists who want to continue working with a messy old medium called paper. Increasingly, if something isn't represented digitally, people will assume it didn't happen.

And although children will clearly use computers at younger and younger ages, we will still need to exercise our analog bodies. Even if she has a computer, a 2-year-old will still want and need to play with her toys and her finger paints. Even if telepresence catches on and we can interact over long distances, we will still want to touch, taste, and smell the people and things we love. As digital bandwidth grows, it will make possible many wonders, but it will never be a substitute for our analog humanity.

Glossary

ADSL See **Asymmetric Digital Subscriber Line**.

Advanced Mobile Phone System The traditional analog cellular phone system.

amplitude modulation A method for encoding radio transmissions using a carrier wave of constant frequency with varying amplitude.

AMPS See **Advanced Mobile Phone System**.

ASCII American Standard Code for Information Interchange.

Asymmetric Digital Subscriber Line A method for using standard telephone lines to transmit data at relatively high speeds.

ATSC Advanced Television Systems Committee.

bandwidth A measure of the information that can flow from one place to another in a given amount of time.

baseband A signal that is transmitted in its final form, often without modulation.

baud A nineteenth-century telegraphy term describing data transmission speed that was used in specifying the **bandwidth** of early modems. Strictly speaking, one baud is one signal event or modulation change per second, but the term is sometimes used incorrectly to mean "characters per second."

broadband A signal that is transmitted in modulated form along with many others in the same medium (over the air or a coaxial cable).

buffer An area of memory used to store extra data temporarily in an attempt to speed or smooth the flow of information.

caching The technique of storing data closer to where it is needed in order to speed retrieval.

carrier frequency A signal that vibrates at a fixed number of cycles per second, or hertz, and is modulated in either frequency or amplitude to carry intelligible information.

carrier wave A radio wave whose amplitude or frequency is varied, or modulated, to encode information. See also **modulation**.

CATV Community Antenna Television.

CDMA See **Code Division Multiple Access**.

CDPD See **Cellular Digital Packet Data**.

Cellular Digital Packet Data A method for transmitting digital data over cellular phones.

CEMA Consumer Electronics Manufacturers Association.

channel A specific carrier frequency.

checksum A number used to test data for the presence of errors that can occur when data is transmitted. The checksum is calculated for a given chunk of data by the sequential combining of all the bytes of data through a series of arithmetic or logical operations.

chrominance The portion of the television signal that contains color information.

Code Division Multiple Access One of several competing digital cellular phone technologies.

crosstalk Leakage of an undesired signal into a desired signal, typically through a wire. See also **interference.**

datacasting The transmission of digital information within a television signal.

dB Decibel.

decibel One-tenth (deci) of a bel, named for Alexander Graham Bell. The decibel scale is logarithmic.

Digital Subscriber Line A generic term for several similar methods of using standard telephone lines to transmit data at relatively high speeds.

digitizing The process of transforming an analog signal into a digital signal. See also **sampling**.

downlink The process of sending information from a satellite down to an earth station or antenna.

DSL See **Digital Subscriber Line**.

dynamic range The range from the loudest to the softest sounds a system can produce.

EDS extended data services.

extended data services A term used for data transmitted in Line 21 of the even-line field of an NTSC television signal.

flat response Absolute accuracy in the reproduction of a signal.

flutter Audible distortion, typically high in pitch, that results from uneven speeds during recording and playing back on analog devices. See also **wow**.

flyback The interval during which the electron beam travels back to the beginning of the next line or field.

forward error correction A means of controlling errors through the sending of redundant information in advance, so that the receiver can calculate the correct data despite normal lapses in the communications channel.

frequency modulation A method for encoding information in an electrical signal by varying its frequency. The FM radio band and the audio portion of broadcast television both use frequency modulation.

granularity The ability to deliver distinct digital data to finer and finer physical locations—ultimately an individual office or desk.

Groupe Special Mobile A form of TDMA technology widely used around the world for digital cellular phones.

GSM See **Groupe Special Mobile**.

harmonic distortion The presence of frequencies in an output signal that were not present in the input, usually expressed as a percentage.

HDTV High-definition television.

hertz (Hz) One cycle per second; the unit is named for Heinrich Hertz.

interference Leakage of an undesired signal into a desired signal, such as interference between two radio stations. See also **crosstalk**.

interlaced scanning Transmitting video by using alternating fields consisting of the picture's odd and even scan lines to minimize bandwidth while avoiding flicker.

International Telecommunication Union The governing body of telecommunications standards.

ISDN Integrated Services Digital Network.

ITU See **International Telecommunication Union**.

latency Delay before the transmission and/or reception of data begins.

lossless compression A compression method that allows the compressed file to be reconstituted in its entirety, bit for bit.

lossy compression A compression method that results in the loss of some data bits when the file is reconstituted.

luminance Amplitude (strength) of the gray scale portion of the television signal.

MMDS Multichannel multipoint distribution system.

Mobile Telephone Switching Office The facility that controls a cellular phone provider's base stations.

modem Short for MOdulator/DEModulator; a communications device that enables a computer to transmit information over a standard telephone line by modulating a computer's digital signals onto a continuous carrier frequency on the telephone line.

modulation The intentional alteration of a signal, usually for the purposes of transmitting information.

MTSO See **Mobile Telephone Switching Office**.

NABTS North American Basic Teletext Specification.

National Television Standards Committee A color television standard consisting of 525 lines scanned at a rate of 60 fields or 30 frames per second.

NTSC See **National Television Standards Committee**.

overscan A portion of the transmitted television image that extends beyond the area of standard television screens.

packet The smallest unit of data transmitted on data networks such as the Internet. Every packet consists of at least two parts: the data and the header.

PAL See **Phase Alternating Line.**

PCN Personal Communication Network; one of several digital cellular phone schemes.

PCS Personal Communication Services; a generic term used to refer to several competing digital cellular phone schemes.

peering The process whereby Internet service providers agree to pass each others' packets along for free.

Permanent Virtual Circuit A method used with frame relay connections to provide an apparent leased line.

Phase Alternating Line A color television standard consisting of 625 lines scanned at a rate of 50 fields or 25 frames per second.

POTS Plain Old Telephone Service.

progressive scanning Transmitting every line of a video picture sequentially. This technique is not used in any broadcast television today but is proposed for future television standards. Nearly all computer displays use progressive scanning.

PVC See **Permanent Virtual Circuit**.

RBDS Radio Broadcast Data System.

roaming The act of using a cell phone outside its local registered area.

routers Devices that receive transmitted packets and forward them to their correct destinations over the most efficient route.

sampling The process of converting an analog signal to digital form by recording the state of the analog waveform (taking a sample) at regular and frequent periods.

sampling depth The number of bits that characterize a single sample.

sampling rate How frequently an analog waveform is analyzed and converted to digital form. To capture an analog waveform, the rate must be at least twice the highest frequency of interest.

SAP Secondary Audio Program.

SDTV Standard-Definition Television.

SECAM See **Séquentiel Coleur Avec Mémoire**.

Séquentiel Coleur Avec Mémoire Sequential color with memory, a television standard consisting of 625 lines scanned at a rate of 50 fields or 25 frames per second.

signal-to-noise ratio The ratio of the desired signal (music, voice, picture) to the noise inherent in any communications channel or medium.

SNR See **signal-to-noise ratio**.

synchronous Sent in real time.

TDMA See **Time Division Multiple Access**.

throughput A measure of data transfer. The useful bandwidth or amount of meaningful data you can transmit, as opposed to the raw bandwidth.

Time Division Multiple Access One of several competing digital cellular phone standards.

UHF Ultra High Frequency band, channels 14–83.

uplink The process of sending information from an earth station or antenna up to a satellite.

VBI See **vertical blanking interval**.

vertical blanking interval The black bars between fields of video, in which data may be transmitted.

VHF Very High Frequency band, channels 2–13.

wow Audible distortion, typically low in pitch, that results from uneven speeds during recording and playing back on analog devices. See also **flutter**.

xDSL A generic term for any of several types of **Digital Subscriber Line**.

Index

Symbols

A

B

C

D

E

F

G

H

I

K

L

M

N

O

P

Q

R

S

T

U

V

W

X

Z

Cary Lu

Born in Qingdao, China, December 4, 1945, Cary Lu was three years old when he came to the United States with his parents. He grew up in California and received an A.B. in physics from the University of California, Berkeley, and a Ph.D in biology from the California Institute of Technology. He also did research in visual perception at Bell Telephone Laboratories.

Cary worked in television for several years on projects for NBC and CBS News, developed short films for Sesame Street and other children's programs on the Public Broadcasting Service, and was the science and technology editor for Children's Television Workshop. He was part of the group that started the Nova series for PBS and he worked on science and technology education projects for the governments of Australia, Kenya, and Algeria. In recent years he was a member of the Committee on Public Understanding of Science and Technology of the American Association for the Advancement of Science.

Cary was founding managing editor of *High Technology* magazine, technology editor for *Inc.* magazine and a columnist on future technology for *Inc. Technology* magazine, but he was best known for his writings on computers. His *The Apple Macintosh Book*, first published in 1984, translated into many languages, and revised in three subsequent editions, was known as the essential volume about that machine. His columns and articles in *Macworld* magazine and other computer and technology publications came to be relied on as models of clarity and integrity.

In person and in print, Cary's ability to explain science and technology to children and adults was a singular gift, and he gave his time selflessly to anyone who asked his assistance. He spent many hours helping with computers at his children's public elementary school, and he was the person his friends came to for definitive answers to complex technological questions.

Cary died of cancer September 23, 1997. He worked on this book right up to the end.

The manuscript for this book was prepared and submitted to Microsoft Press in electronic form. Text files were prepared using Microsoft Word for Windows. Pages were composed by Studioserv (www. studioserv.com) using Adobe PageMaker 6.5 for Windows, with text in Optima and display type in Optima Bold. Composed pages were delivered to the printer as electronic prepress files.

Cover Graphic Designer and Illustrator
Becker Design

Interior Graphic Designer
Kim Eggleston

Interior Graphic Artist
Travis Beaven

Manuscript Editor
Gail Taylor

Principal Compositor
Sharon Bell

Project Coordinator and Technical Editor
Devra Hall

Proofreader
Tom Speeches

Indexer
Audrey Marr

Register Today!

Return this
The Race for Bandwidth:
Understanding Data Transmission
registration card for
a Microsoft Press® catalog

U.S. and Canada addresses only. Fill in information below and mail postage-free.
Please mail only the bottom half of this page.

1-57231-513-X ***THE RACE FOR BANDWIDTH: UNDERSTANDING DATA TRANSMISSION*** ***Owner Registration Card***

____
NAME

____
INSTITUTION OR COMPANY NAME

____
ADDRESS

____

____
CITY STATE ZIP

NO POSTAGE
NECESSARY
IF MAILED
IN THE
UNITED STATES

BUSINESS REPLY MAIL
FIRST-CLASS MAIL PERMIT NO. 53 BOTHELL, WA

POSTAGE WILL BE PAID BY ADDRESSEE

MICROSOFT PRESS REGISTRATION
THE RACE FOR BANDWIDTH:
UNDERSTANDING DATA TRANSMISSION
PO BOX 3019
BOTHELL WA 98041-9946